nuclear Magnetic
Resonance
Shift Reagents

ACADEMIC PRESS RAPID MANUSCRIPT REPRODUCTION

These writings are based in part on a Symposium on the Chemistry of Nuclear Magnetic Resonance Shift Reagents, sponsored by the Analytical and Organic Chemistry Divisions of the American Chemical Society, at the 165th National A. C. S. Meeting, Dallas, Texas, April 9-11, 1973

nuclear Magnetic Resonance Shift Reagents

Edited by

Robert E. Sievers

Aerospace Research Laboratories
Wright-Patterson Air Force Base, Ohio

ACADEMIC PRESS New York and London 1973

A Subsidiary of Harcourt Brace Jovanovich, Publishers

ACADEMIC PRESS, INC.
111 Fifth Avenue, New York, New York 10003

United Kingdom Edition published by
ACADEMIC PRESS, INC. (LONDON) LTD.
24/28 Oval Road, London NW1

LIBRARY OF CONGRESS CATALOG CARD NUMBER: 72-12205

PRINTED IN THE UNITED STATES OF AMERICA

CONTENTS

CONTENTS

CONTRIBUTORS

Ian M. Armitage, Department of Chemistry, University of British Columbia, Vancouver 8, Canada

Joachim Bargon, IBM Research Laboratory, Molecular Physics Department, San Jose, California 95193

C. D. Barry, Department of Biophysics, University of Oxford, Oxford, England

F. Behbahany, Research Center, National Iranian Oil Company, Tehran, Iran

W. A. Boyd, Department of Chemistry and Biochemistry, Southern Illinois University, Carbondale, Illinois 62901

J. M. Briggs, Department of Chemistry, Queen Mary College, Mile End Road, London E1 4NS, England

J. J. Brooks, Aerospace Research Laboratories, ARL/LJ, Wright-Patterson AFB, Ohio 45433

A. H. Bruder, Department of Chemistry, State University of New York, Stony Brook, New York 11790

R. A. Cooper, Gates and Crellin Laboratories of Chemistry, California Institute of Technology, Pasadena, California 91109

J. A. Cunningham, Aerospace Research Laboratories, ARL/LJ, Wright-Patterson AFB, Ohio 45433

Raymond E. Davis, Department of Chemistry, University of Texas, Austin, Texas 78712

C. M. Dobson, Department of Inorganic Chemistry, University of Oxford, Oxford, England

D. S. Dyer, Aerospace Research Laboratories, ARL/LJ, Wright-Patterson AFB, Ohio 45433

L. E. Ford, Department of Biophysics, University of Oxford, Oxford, England

A. D. Godwin, Department of Chemistry, Texas A&M University, College Station, Texas 77843

Laurance D. Hall, Department of Chemistry, University of British Columbia, Vancouver 8, Canada

F. A. Hart, Department of Chemistry, Queen Mary College, Mile End Road, London E1 4NS, England

G. E. Hawkes, Gates and Crellin Laboratories of Chemistry, California Institute of Technology, Pasadena, California 91109

K. Herwig, Gates and Crellin Laboratories of Chemistry, California Institute of Technology, Pasadena, California 91109

C. C. Hinckley, Department of Chemistry and Biochemistry, Southern Illinois University, Carbondale, Illinois 62901

William DeW. Horrocks, Jr., Department of Chemistry, The Pennsylvania State University, University Park, Pennsylvania 16802

S. R. Johns, Gates and Crellin Laboratories of Chemistry, California Institute of Technology, Pasadena, California 91109

M. D. Johnston, Jr., Department of Chemistry, Texas A&M University, College Station, Texas 77843

J. S. Kristoff, Department of Chemistry, Northwestern University, Evanston, Illinois 60201

Charles Kutal, Department of Chemistry, University of Georgia, Athens, Georgia 30602

D. Leibfritz, Gates and Crellin Laboratories of Chemistry, California Institute of Technology, Pasadena, California 91109

R. Burton Lewis*, Department of Chemistry, Indiana University, Bloomington, Indiana 47401

T. J. Marks, Department of Chemistry, Northwestern University, Evanston, Illinois 60201

*Present address: Western Electric, Box 900, Princeton, New Jersey 08540

Alan G. Marshall, Department of Chemistry, University of British Columbia, Vancouver 8, Canada

C. Marzin, Gates and Crellin Laboratories of Chemistry, California Institute of Technology, Pasadena, California 91109

G. P. Moss, Department of Chemistry, Queen Mary College, Mile End Road, London E1 4NS, England

H. L. Pearce, Department of Chemistry, Texas A&M University, College Station, Texas 77843

M. Pickering, Department of Chemistry, State University of New York, Stony Brook, New York 11790

R. Porter, Department of Chemistry, Northwestern University, Evanston, Illinois 60201

T. W. Proulx, Department of Chemistry, Texas A&M University, College Station, Texas 77843

E. W. Randall, Department of Chemistry, Queen Mary College, Mile End Road, London E1 4NS, England

Jacques Reuben, Isotope Department, The Weizmann Institute of Science, Rehovot, Israel

D. W. Roberts, Gates and Crellin Laboratories of Chemistry, California Institute of Technology, Pasadena, California 91109

J. D. Roberts, Gates and Crellin Laboratories of Chemistry, California Institute of Technology, Pasadena, California 91109

H. A. Rockefeller, Department of Chemistry, State University of New York, Stony Brook, New York 11790

R. E. Rondeau, Air Force Materials Laboratory, AFML/LPH, Wright-Patterson AFB, Ohio 45433

K. D. Sales, Department of Chemistry, Queen Mary College, Mile End Road, London E1 4NS, England

B. L. Shapiro, Department of Chemistry, Texas A&M University, College Station, Texas 77843

M. J. Shapiro, Department of Chemistry, Texas A&M University, College Station, Texas 77843

D. F. Shriver, Department of Chemistry, Northwestern University, Evanston, Illinois 60201

R. E. Sievers, Aerospace Research Laboratories, ARL/LJ, Wright-Patterson AFB, Ohio 45433

James P. Sipe, III, Department of Chemistry, The Pennsylvania State University, University Park, Pennsylvania 16802

G. V. Smith, Department of Chemistry and Biochemistry, Southern Illinois University, Carbondale, Illinois 62901

C. S. Springer, Department of Chemistry, State University of New York, Stony Brook, New York 11790

M. L. Staniforth, Department of Chemistry, Queen Mary College, Mile End Road, London E1 4NS, England

Daniel Sudnick, Department of Chemistry, The Pennsylvania State University, University Park, Pennsylvania 16802

D. A. Sweigart, Department of Chemistry, Swarthmore College, Swarthmore, Pennsylvania 19081

S. R. Tanny, Department of Chemistry, State University of New York, Stony Brook, New York 11790

R. L. R. Towns, Department of Chemistry, Texas A&M University, College Station, Texas 77843

Ernest Wenkert, Department of Chemistry, Indiana University, Bloomington, Indiana 47401

Lawrence G. Werbelow, Department of Chemistry, University of British Columbia, Vancouver 8, Canada

M. Robert Willcott, III, Department of Chemistry, University of Houston, Houston, Texas 77004

R. J. P. Williams, Department of Inorganic Chemistry, University of Oxford, Oxford, England

PREFACE

Seldom has a field grown as explosively as the chemistry of lanthanide shift reagents. The subject has mushroomed from a handful of papers in the 1960's to more than 400 publications in the last four years related in one way or another to the chemistry and use of shift reagents. In preparing this volume an attempt was made to describe the progress that has been made in elucidating the phenomena and present some of the most recent findings of research groups active in this field.

Clearly, the use of shift reagents is an idea whose time has come. The great need for a simple aid to the chemist for easing the difficulty of interpretation of complex spectra and enabling him to confirm assignments unequivocally has provided much of the impetus in this field. More sophisticated applications such as structure and conformation studies in solution, measurements of optical purity, studies of isotope effects, studies of biologically important molecules and organometallic compounds, and the like are just beginning to blossom, and the latest developments are treated in the various chapters of this book.

In a field that is growing so rapidly it is inevitable that many honestly held disagreements will occur between researchers. In time our knowledge will doubtless be perfected, but some of these differences are necessarily reflected in this book and the extensive literature listed in the Bibliography. Orthodoxy is unattainable at this point and may never be, but at least there is a full and open discussion of the evidence and reasoning.

Since the literature is so voluminous the Bibliography at the end of the book is divided into the following sections to aid the reader:

1. Fundamental Aspects of Shift Reagent Chemistry
2. Chemical and Physical Properties of Shift Reagents
3. X-ray Crystallographic Structural Determinations
4. General Applications of Shift Reagents
5. Applications of Shift Reagents to Studies of Biologically Significant Molecules
6. Determination of Enantiomeric and Diastereomeric Compositions and Related Phenomena
7. Kinetic Studies Involving Shift Reagents

8. Isotope Effects in Shift Reagent Chemistry
9. Applications of Shift Reagents to the Study of Stereochemistry

For the many articles which treat several subjects, the choice of subtopic listing is somewhat arbitrary, so other sections should be consulted if a paper is not found immediately.

The symbols used in each chapter are defined as they occur, but a glossary of abbreviations for compounds is included at the end of the book since many of the shift reagents are often referred to by their trivial names or abbreviations (*e.g.*, the dipivaloylmethanates, dpm) rather than the IUPAC systematic names.

I am grateful to several of my co-workers, in particular Drs. J. J. Brooks, J. A. Cunningham, D. S. Dyer and C. Kutal, who contributed much of their own time and energies to this undertaking. I also extend special thanks to Mrs. Inez Oloff and Mrs. Babs Brooks, who provided very effective secretarial assistance. To my family, for their patience and support while I was inaccessible even more than ordinarily, I can only say that I hope to make up for those lost times.

Robert E. Sievers

nuclear Magnetic Resonance Shift Reagents

Chemistry of Lanthanide Shift Reagents
Secondary Deuterium Isotope Effects

C. C. Hinckley, W. A. Boyd, and G. V. Smith

Department of Chemistry and Biochemistry
Southern Illinois University
Carbondale, Illinois   62901

and

F. Behbahany

Research Center
National Iranian Oil Company
Tehran, Iran

Paramagnetic lanthanide β-diketoenolate complexes have been found useful in high resolution nuclear magnetic resonance (nmr) studies of organic compounds.[1,2,3] In solution, lanthanide complexes associate with basic functional groups of organic compounds. Magnetic interactions accompanying the association induce shifts in nmr spectra of the organic substrates. Metal complexes useful in this context have been called "shift reagents". With lanthanides the predominant magnetic interaction is pseudocontact in nature[4], Eq. (1).

$$\left(\frac{\delta}{\nu}\right)_i = K_{axial}\left[\frac{3\cos^2\Theta_i - 1}{R_i^3}\right] + K_{nonax}\left[\frac{\sin^2\Theta_i \cos2\emptyset}{R_i^3}\right] \quad (1)$$

When europium and praseodymium complexes are used, the shifts, δ_i, are accompanied with little broadening, therefore, the identification of the shifted resonances is possible. Applications include the simplified analysis of nmr spectra by separation of overlapping resonances and sub-

1

sequent assignment of these resonances. The data
provided by the nature of the magnetic interaction
are perhaps of greater importance. Since this
interaction depends upon parameters of molecular
structure (distance, R, and angles θ and ϕ) pseudo-
contact shift data may be used to study molecular
structure in solution.

The principal complexes used have been the
europium and praseodymium chelates of dipivalo-
methane[5] (HDPM, R=R'=t-butyl) and 6,6,7,7,8,8,8-
heptafluoro-2,2-dimethyl-3,5-octanedione[6] (Hfod,
R=t-butyl and R'=perfluoropropyl),I.

$$3 \left[Ln \underset{O == C}{\overset{O --- C}{\diagdown}} \begin{matrix} R' \\ \diagup \\ \diagdown CH \\ \diagdown \\ R \end{matrix} \right] \qquad I$$

These compounds associate with a wide variety of
basic substrates by readily expanding their coordination numbers beyond six. Europium compounds
generally induce downfield shifts in substrate
resonances, while the praseodymium analogs generally cause upfield shifts.

The chemistry of these systems is complex.
Both 1:1 and 2:1 substrate to metal chelate stoi-
cheometries are found.[1,2,3] Dimers (polymers) of
the metal chelates are found in the solid[7] and in
solution[8]. A class of phenomena, best collected
together as the "impurity problem", have been
observed which are only marginally explained.
Reliable determinations of what equilibria are
present are needed, and accurate values of all
the appropriate equilibrium constants for these
systems are of more than passing interest. The
relation between the observed lanthanide induced
nmr and the chemistry of these systems is intimate.
A major direction of study in the area of lanthanide shift reagents is the development of pseudo-
contact shift analyses as a reliable tool in
structure determination. As these efforts progress beyond model compounds to substrates of substantially unknown structure, confident knowledge
of the species in solution will become increasing-

ly important.

In this paper the chemistry of lanthanide shift reagents is discussed from the points of view of several different models. The effort has been made to incorporate several aspects of shift reagent chemistry. A graphical presentation useful in visualizing interrelationships between the several species has been adopted[9]. The graphs are constructed from metal complex saturation factors on which observed lanthanide induced shifts depend. Competition between substrates is discussed. Deuterium isotope effects observed in lanthanide induced shifts are reviewed as a special case of competition.

Graphical Representations

The first and simplest model of LSR chemistry to be quantitatively applied to lanthanide induced shifts was 1:1 metal chelate, M, substrate, S, complex formation as,

$$M \cdot S \rightleftharpoons M + S \qquad K_1 = \frac{[M][S]}{[M \cdot S]} \qquad (2)$$

In this case the induced shift, Δ, with a ligand exchange, is given for the ith resonance by,

$$\Delta_i = \frac{[M \cdot S]}{S_T} \delta_{1_i} \qquad (3)$$

where $[M \cdot S]$ is the concentration of the complex $M \cdot S$, S_T is the total substrate concentration, and δ_{1_i} is the limiting shift obtained from Eq. (1). Using the definitions,

$$\rho = \frac{M_T}{S_T} \qquad (4) \qquad \text{and} \qquad \alpha_1 = \frac{[M \cdot S]}{M_T} \qquad (5)$$

where M_T is total metal complex concentration, Eq. (3) becomes

$$\Delta_i = \alpha_1 \rho \delta_{1_i} \qquad (6)$$

The induced shift is predicted to be a linear function of the variable ρ. Experimentally, linearity is often found in plots of Δ vs. ρ when

3

$\rho < 0.7$, and the shift reagent studied is Eu(DPM)$_3$[10]. In a typical study lanthanide chelate is added in small solid increments to a solution of substrate in a suitable solvent and nmr spectra obtained. Frequency positions, or alternatively induced shifts, of nuclear resonances are plotted versus ρ. Figure 1 is typical of the results obtained. Since δ_{1_i} is constant and ρ is an independent variable, the evident curvature of the plot is a reflection of the function α_1 (Eq. (6)). The saturation factor, α_1, is dependent upon the concentration of free substrate, [S], and the dissociation constant, K_1, and is given by,

$$\alpha_1 = \frac{[M \cdot S]}{M_T} = \frac{[S]}{[S] + K_1} \tag{7}$$

Similarly,

$$\alpha_0 = \frac{[M]}{M_T} = \frac{K_1}{[S] + K_1} \tag{8}$$

Because of the intimate relationship between the observed shifts in LSR experiments and the metal complex saturation factors, graphs based on these functions are a useful adjunct in correlating the several elements involved. Figure 2 is a log-log plot of species concentration versus the free substrate concentration for a 1:1 system which is based on the metal saturation factors, α_1 and α_0. All species are represented. The point [log K_1^0, log M_T] is the "system point" and plots of all metal species turn on that point. The graph represents solutions of metal concentration M_T over a wide range of substrate concentrations. Relative concentrations for a particular case may be obtained from the graph by the application of suitable mass-balance conditions. For instance, when $\rho = 1$, $M_T = S_T$, and mass-balance requires that

$$[M] = [S]. \tag{9}$$

This condition is satisfied in the graph in Figure 2 at the intersection of the log [M] and log

4

[S] plots. A more general relation for ρ-values other than unity is

$$[M] = [S] + C \qquad (10)$$

where C is given by

$$M_T = S_T + C \qquad (11)$$

Figure 2 is one graph of a series, each for a different M_T and associated mass balance relations, needed to deal with variations of both metal and substrate concentrations. With the aid of this graph it is possible to make some general comments about 1:1 systems.

For solutions having compositions which fall to the right of the system point, essentially all of the metal complex is associated with substrate. In this region, α_1 is at its maximum value with the result that plots of Δ versus ρ are linear. For any solution composition, ratios of observed resonance shifts are equal to the ratios of limiting shifts, δ_{1_i}. In principle, the limiting shift is never observed, but in practice may be approached when ρ is large and the dissociation constant, K_1, is small.

The graph in Figure 2 may be applied to many shift reagent systems involving $Eu(DPM)_3$ and the value of $K_1 = 10^{-3} \underline{M}$ is on the order of those observed for alcoholic substrates. The more basic the substrate the smaller is the value of K_1. In graphical terms, the system point moves to the left with increasing basicity of the substrate.

A more inclusive model of shift reagent chemistry includes 2:1 association. The equilibria are,

$$M \cdot S_2 \rightleftharpoons M \cdot S + S \qquad K_2 = \frac{[M \cdot S][S]}{[M \cdot S_2]} \qquad (12)$$

$$M \cdot S \rightleftharpoons M + S \qquad K_1 = \frac{[M][S]}{[MS]} \qquad (13)$$

and expressions for the associated saturation factors are,

$$\alpha_2 = \frac{[M \cdot S_2]}{M_T} = \frac{[S]^2}{[S]^2 + [S]K_2 + K_1K_2} \qquad (14)$$

$$\alpha_1 = \frac{[M \cdot S]}{M_T} = \frac{[S]K_2}{[S]^2 + [S]K_2 + K_1K_2} \qquad (15)$$

$$\alpha_0 = \frac{[M]}{M_T} = \frac{K_1K_2}{[S]^2 + [S]K_2 + K_1K_2} \qquad (16)$$

Figure 3 is a log-log plot of species concentration versus [S] based on these equations. There are two system points. In this case, the observed shifts, Δ_i, are given by

$$\Delta_i = \alpha_1 \rho \delta_{1_i} + 2\alpha_2 \rho \delta_{2_i} \qquad (17)$$

where α_1 and α_2 are the saturation factors for the species $M \cdot S$ and $M \cdot S_2$. δ_{1_i} and δ_{2_i} are the limiting shifts of the ith resonance in the two complexes. These shifts must in principle be considered as different and are in fact found to be different. Both complexes, $M \cdot S$ and $M \cdot S_2$, may be present simultaneously (Figure 4). This reality complicates the analysis considerably. In the simple 1:1 association model the issues are comparatively straight forward. The questions there are concerned with the value of K_1 and solubility. In double association, these problems are amplified by the need to know the value of two dissociation constants and the relation between them. The immediate problem faced by the experimentalist is the fact that the observed shift is no longer directly related to a single limiting shift but is an average of two such quantities.

The compounds now in use as shift reagents, $Ln(DPM)_3$ and $Ln(fod)_3$, exhibit the full range of

6

single and double association. Double association is a significant element in the chemistry of
Ln(DPM)$_3$ chelates. Eu(DPM)$_3$·2py is obtained as
a crystaline solid. Single association for both
Eu(DPM)$_3$ and Pr(DPM)$_3$ seems to be the rule, however, when alcohols or compounds less basic than
alcohols are the substrates. Graphically (Figure
3) this means that for these metal chelate-substrate combinations, the system point associated
with K$_2$ is far to the right. In the case of the
Ln(fod)$_3$ compounds, double association is much
more important. Shapiro and Johnston have developed a method for determining both K$_1$ and K$_2$ for
Eu(fod)$_3$ systems[11].

The Dimer

Dimeric forms of the compounds Ln(DPM)$_3$ and
Ln(fod)$_3$ are known to be present in the crystalline state[7] and there is substantial evidence
that equilibria involving dimers are important
in solution as well. Desreux and coworkers have
shown through vapor phase osmometry that dimers
of Eu(fod)$_3$ predominate in non-associative solvents and have found evidence for trimers[8]. Including the dimer, solution equilibria in this
third model become,

$$M \cdot S_2 \rightleftharpoons M \cdot S + S \qquad K_2 = \frac{[M \cdot S][S]}{[M \cdot S_2]} \qquad (18)$$

$$M \cdot S \rightleftharpoons M + S \qquad K_1 = \frac{[M][S]}{[M \cdot S]} \qquad (19)$$

$$2M \rightleftharpoons M_2 \qquad K_0 = \frac{[M_2]}{[M]^2} \qquad (20)$$

Saturation factors are,

$$\alpha_2 = \frac{[S]^2}{Q} \qquad (21)$$

$$\alpha_1 = \frac{K_2[S]}{Q} \qquad (22)$$

$$\alpha_0 = \frac{K_1 K_2}{Q} \tag{23}$$

$$\alpha_D = \frac{\sqrt{A^2+B} - A}{\sqrt{A^2+B} + A} \tag{24}$$

where,

$$A = [S]^2 + K_2[S] + K_1 K_2 \tag{25}$$

$$B = 4K_1^2 K_2^2 K_D M_T \tag{26}$$

$$Q = \left(1/2\right)\left(A + \sqrt{A^2 + B}\right) \tag{27}$$

The above relationships are substantially more involved than any of those preceeding. It is, however, interesting that the equation for the observed shift is still Equation 17. That is, the complexes responsible for the shifts remain M·S and M·S$_2$. No new substrate-metal species have been introduced. The dimer equilibrium simply competes with the substrate. These factors are not insignificant. All studies, so far, of shift reagent equilibria which have attempted to calculate the dissociation (association) constants involved have been based on analyses of substrate resonance shifts only. No attempt has been made to include shifts observed in the metal complex resonances. Furthermore, investigators have not been compelled by their data to include dimeric species. Dissociation (association) constants have been obtained with apparently satisfactory accuracy ($\sim$5%)[10,11]. A question naturally arises. If a dimer species is involved, and the evidence for it is compelling, why is it not more obvious in analyses of observed substrate resonance shifts? The graphical representation of the system suggests an answer. Figure 5 is a log-log plot of species concentration versus free substrate concentration. The values of K_1 and K_2 are chosen to be the same as those of Figure 3. K_D=100, a value similar to that determined by Desreux and coworkers for the solvent carbon tetrachloride.

The expanded system is obviously more complicated than simple double association. Free metal
complex, M, is the principal species at no point
on the graph. In spite of these differences, however, the general similarity with the system of
Figure 3 is striking. There are still two apparent system points and the substrate concentra-
tion dependence of the species $M \cdot S$ and $M \cdot S_2$ is
virtually just as before. A telling feature of
the graph is that the system point formerly
associated with K_1 has been translated to the
right, toward less association. The apparent
dissociation constant, K', for this relocated
system point is given by,

$$K' = (1/2) \ K_1 [1 + \sqrt{1 + 8K_D M_T} \ ] \tag{28}$$

Determinations of the constants of such a
system, from substrate resonance shifts only, with
out explicitly including the dimer equilibrium,
will yield K_2 and K'. Metal concentration dependence of K' is suppressed by the square root, and
so K' may appear to be K_1 .

Competition

The principal effect of the dimer is one of
competition with the substrate. Dimer participation, however, is not established for all shift
reagents. $Pr(DPM)_3$ and $Eu(DPM)_3$, though dimeric
in the solid, may be principally monomer in solution. Whether the dimer is involved or not, competition is a factor that must be considered often.
Reuben[3] and others[10] have examined the effects of
impurities and have established that small quantities of some competitors may affect observed
shifts.

A simple model for competition based on 1:1
association may be written

$$M \cdot S \rightleftharpoons M + S \qquad K_{1S} = \frac{[M][S]}{[M \cdot S]} \tag{29}$$

$$M \cdot P \rightleftharpoons M + P \qquad K_{1P} = \frac{[M][P]}{[M \cdot P]} \tag{30}$$

S and P are the competing substrates. Saturation factors are given by,

$$\alpha_{1S} = \frac{[M \cdot S]}{M_T} = \frac{K_{1P}[S]}{K_{1P}[S] + K_{1S}[P] + K_{1P}K_{1S}} \qquad (31)$$

$$\alpha_{1P} = \frac{[M \cdot P]}{M_T} = \frac{K_{1S}[P]}{K_{1P}[S] + K_{1S}[P] + K_{1P}K_{1S}} \qquad (32)$$

$$\alpha_0 = \frac{[M]}{M_T} = \frac{K_{1P}K_{1S}}{K_{1P}[S] + K_{1S}[P] + K_{1P}K_{1S}} \qquad (33)$$

In a case where an impurity is present at a constant concentration, say $[P]_o$, determination of the dissociation constant for the substrate S, will yield an apparent constant K'_S, where

$$K'_S = K_{1S}[1 + \frac{[P]_o}{K_{1P}}] \qquad (34)$$

Thus, the effect of competition in this case is similar to that found for the dimer.

Competition between substrates when double association is considered is, of course, elaborate. It is necessary in this case to introduce a new species, M·SP, the mixed adduct.

$$M \cdot SP \rightleftharpoons M \cdot S + P \qquad K_{SP} = \frac{[M \cdot S][P]}{[M \cdot SP]} \qquad (35)$$

$$M \cdot SP \rightleftharpoons M \cdot P + S \qquad K_{PS} = \frac{[M \cdot P][S]}{[M \cdot SP]} \qquad (36)$$

Though the species M·SP is involved in two separate equilibria it can be shown that K_{SP} and K_{PS} are not independent. The importance of the mixed adduct will vary from case to case but it should be considered part of the system. Some of the early work in lanthanide shift reagent studies[1] was with the complex $Eu(DPM)_3 \cdot 2py$ (py=pyridine). Pyridine in this case is a constant competitor, and studies of these systems have indicated that the mixed adduct is of primary significance.

Isotope Effect

The deuterium isotope effect in lanthanide shifts is a special case of competition. Deuterium substitution near (usually geminal) the basic coordination site of the substrate increases the basicity of the substrate. The effect is readily observed in 1:1 mixtures of deuterium substituted (labeled) and unsubstituted (unlabeled) alcohols as a doubling of all resonance peaks concomitant with the shifts induced by the lanthanide complex.[12,13] The effect has also been observed with a suitably labeled aldehyde, and amines and ethers. Peaks associated with the labeled compound are shifted further, indicating that deuterium substitution decreases the dissociation constant of the metal chelate-substrate complex. Differences in shifts are on the order of 2-3 percent (Table 1).

Sunko and Borcic have recently reviewed work in the area of secondary deuterium isotope effects.[14] A secondary deuterium isotope effect is one in which the bond to the deuterium atom is not broken in the rate determining step. In solvolyses, deuterium substitution on the carbon at the reaction site (α-effect) normally slows the solvolysis reaction. The change in reaction rate is associated to changes in the transition state due to the isotopic substitution. Effects presented in this paper are for systems in equilibrium and are not reaction rate effects, however, some general correlations with rate data are possible. If the equilibrium between shift reagent and substrate are described in terms of an association step and a dissociation step, then deuterium substitutions would be expected to have little effect on the association step. If the dissociation step is slowed by deuterium substitution then a smaller overall dissociation constant for the equilibrium would result, which is the finding.

The isotope effect has also been observed when the deuterium substitution is on a carbon atom adjacent to the coordination site (β-effect).

11

Sanders and Williams found differences in lanthanide induced shifts for labeled and unlabeled 1, 1,2 trimethyl propanols in which deuterium substitution were on methyl groups adjacent to the coordinating hydroxyl.[15] We have found that deuterium substitution in 4-picoline also produces differences in shift. In these experiments labeling of the picoline was mixed and precise shift differences were unavailable, but evident shift differences of about 2% were found. Observation of these β-effects indicates that direct participation of the deuterium in bonding between the metal chelate and the substrate, though possible in some cases, is not a primary contributor to the isotope effect.

Differences in the observed isotope effects, measured as percent of shift, for similar substrates indicates dependence upon rather subtle structure factors (Table 1). The isotope effect for verbanol is 1.9% while that for isoverbanol is 1.3% even though the geometries of the two molecules are similar at the coordination site. Patrick and Patrick have seen D-C-O bond angle dependences in the isotope effect for some alcohols.[16] Comparison of the isotope effect for 1-deuterioethanol (1.5%) and 1,1-dideuterioethanol (3.%) shows near additivity for multiple substitution. These data are indicative of the information which may be obtained and demonstrate that this phenomenon is likely to be useful in the study of secondary deuterium isotope effects. The data is, however, not definitive. The chemistry of lanthanide shift reagents is complex and care must be exercised to avoid misinterpretation of the data. For instance, multiple associations must be considered. Percent differences in observed shifts are calculated from the ratio

$$\frac{\Delta_D - \Delta_H}{\Delta_H} = \frac{\Delta_D}{\Delta_H} - 1 \qquad (37)$$

The quantity Δ_D/Δ_H is in principle complex, but it is useful to consider a special case. If only 1:1 association is assumed, if $\rho_H = \rho_D$, and we approxi-

mate equal concentration for the labeled and unlabeled free substrate, then

$$\frac{\Delta_D}{\Delta_H} = \frac{K_H}{K_P} = K_{HD} \tag{38}$$

The ratio of shifts in this case is a direct measure of the equilibrium constant for exchange of
the labeled and unlabeled substrate with the metal chelate complex according to the equation,

$$M \cdot S_D + S_H \xrightleftharpoons{K_{HD}} M \cdot S_H + S_D \tag{39}$$

where S_D and S_H represent the labeled and unlabeled substrate respectively.

Editor's Note: Independent gas chromatographic measurements relating to the secondary deuterium isotope effect
exhibited by shift reagents are discussed in the following
chapter.

References

1. C. C. Hinckley, J. Amer. Chem. Soc., <u>91</u>, 5160 (1969).

2. R. von Ammon and R. D. Fischer, Angew. Chem. Internat. Edit., 675 (1972).

3. J. Reuben, Prog. Nucl. Mag. Res. Spect., in press.

4. B. Bleaney, J. Mag. Res., <u>8</u>, 91 (1972).

5. J. K. M. Sanders and D. H. Williams, Chem. Comm., 422 (1970).

6. R. E. Rondeau and R. E. Sievers, J. Amer. Chem. Soc. <u>93</u>, 1522 (1971).

7. C. S. Erasmus and J. C. A. Boeyers, Acta. Crystallogr. Sec. B., <u>26</u>, 1843 (1970).

8. J. F. Desreux, L. E. Fox, and C. N. Reilley, Anal. Chem., <u>44</u>, 2217 (1972).

9. L. G. Sillen, "Graphical Presentation of Equilibrium Data", Treatise on Anal. Chem., <u>1</u>, pt. 1, Koltoff, ed., 277 (1959).

10. J. K. M. Sanders, S. W. Hanson, and D. H. Williams, J. Amer. Chem. Soc., <u>94</u>, 5325 (1972)

11. B. L. Shapiro and M. D. Johnston, Jr., J. Amer. Chem. Soc., <u>94</u>, 8185 (1972).

12. G. V. Smith, W. A. Boyd, and C. C. Hinckley, J. Amer. Chem. Soc., <u>93</u>, 6319 (1971).

13. C. C. Hinckley, W. A. Boyd, and G. V. Smith, Tetrahed. Lett., 879 (1972).

14. D. E. Sunko and S. Borcic, "Secondary Deuterium Isotope Effects and Neighboring Group Participation", Isotope Effects in Chemical Reactions, D. J. Collins and N. S. Bowman, eds., Van Nostrand Reinhold Company, 1970, pp. 160-212.

15. J. K. M. Sanders and D. H. Williams, Chem. Comm., 436 (1972).

16. T. B. Patrick and P. H. Patrick, private comm.

Table I

Isotope Effects on NMR Shifts Induced by Lanthanide Shift Reagents

Substrate	Lanthanide Complex	% Difference
	$Eu(DPM)_3 \cdot 2py$	2.9
	"	1.5
	"	1.8
	"	1.3
	"	4.3
	"	3.4
	"	2.4
	"	2.7

15

Table I (Cont.)

CH_3 $\overset{H}{\underset{OH}{|}}$ D	"	1.3
CH_3 $\overset{D}{\underset{OH}{|}}$ D	"	2.5

	Eu(DPM)$_3$	2.6
	Pr(DPM)$_3$	2.6
CH_3 $\overset{D}{\underset{OH}{|}}$ CH_2CH_3	Eu(DPM)$_3$	2.2
	Pr(DPM)$_3$	1.7
	Eu(BAT)	2.2
	Eu(DBM)$_3$	2.9
	Eu(DPM)$_3$	~2
$\overset{NH_2}{\underset{D}{|}}$ CH_3 (cyclohexyl)	Eu(DPM)$_3$·2py	2.4
$\overset{NH_2}{\underset{D}{|}}$ CH_3 (cyclohexyl)	"	4.7

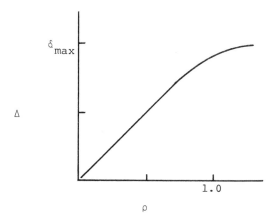

Figure 1. Plot of observed shift, Δ , versus ρ typical of those obtained for 1:1 systems.

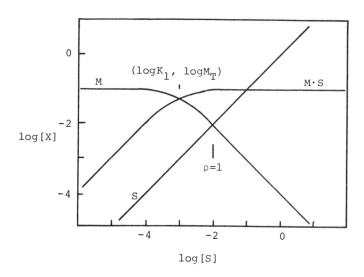

Figure 2. Log-log plot of species concentration, [X], versus free substrate concentration, [S], for a 1:1 system. Total metal concentration, M_T, is 0.1 $\underline{M}$ and $K_1 = 10^{-3}$ $\underline{M}$.

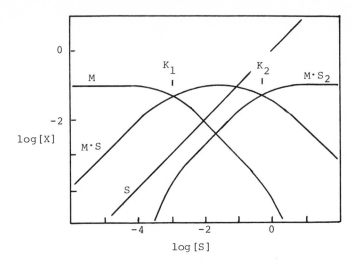

Figure 3. Log-log plot of species concentration versus free substrate concentration for a 2:1 system. $K_1 = 10^{-3}$ M, $K_2 = 10^{-0.3}$M, and $M_T = 0.1$ M.

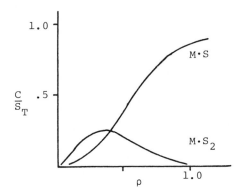

Figure 4. Plots of relative concentrations of M·S and M·S$_2$ for the system in Figure 3.

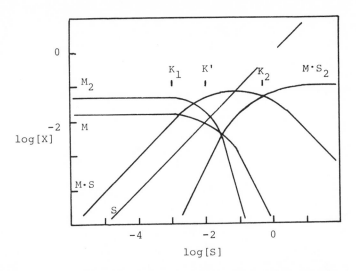

Figure 5. Log-log plot of species concentration versus free substrate concentration for a 2:1 system including metal chelate dimerization. $K_1 = 10^{-3}$ $\underline{M}$, $K_2 = 10^{-0.3}$ $\underline{M}$, $K_D = 10^2$ $\underline{M}^{-1}$ and $M_T = 0.1$ $\underline{M}$.

INTERACTIONS OF NUCLEOPHILES WITH
LANTHANIDE SHIFT REAGENTS

D.S. Dyer, J.A. Cunningham,
J.J. Brooks, and R.E. Sievers
Aerospace Research Laboratories, ARL/LJ
Wright-Patterson AFB, Ohio 45433

R.E. Rondeau
Air Force Materials Laboratory, AFML/LPH
Wright-Patterson AFB, Ohio 45433

INTRODUCTION

Since Hinckley demonstrated the practical application of a lanthanide chelate for inducing shifts in NMR spectra (1) (the LIS phenomenon), NMR shift reagents have experienced increasing popularity particularly in the area of spectral clarification. A large portion of the NMR shift reagent literature devotes relatively little attention to the nature of the shift reagent or the species it forms with the substrate, since the goal of these studies (i.e., the elucidation of complex spectra) can be attained without detailed information concerning the nature of the complex. Considerable effort has been extended on many fronts to more fully explain the physical basis for the LIS phenomenon and therby bring theory up with practice. These studies have provided valuable information in areas ranging from solution configuration determination (2) to the selection of the best shift reagent for a given application (3).

Since the key to the LIS phenomenon is the formation of a complex between a nucleophile and a coordinately unsaturated lanthanide chelate, it is important to understand the nature of this interaction. Of the many techniques for studying interactions between substrates and NMR shift reagents, we have concentrated our efforts in three areas: NMR studies, gas chromatographic studies, and X-ray structural studies.

21

The most frequently used chelate shift reagents are the europium(III), praseodymium(III), and ytterbium(III) chelates of fod, thd (also abbreviated dpm), and facam(4)

$$H_3C-C-C-C-C-CF_2-CF_2-CF_3 \qquad \text{fod}$$

$$H_3C-C-C-C-C-C-CH_3 \qquad \text{thd}$$
$$\text{(or dpm)}$$

facam

The tris lanthanide chelates of thd were first synthesized and characterized by Eisentraut and Sievers (5) in the early 1960's; the fod chelates were studied by Springer, Meek and Sievers (6) and the facam chelates by Feibush, et al. (7,8). In these and subsequent studies we designed lanthanide chelates that are highly soluble in non-polar organic solvents and can function effectively as Lewis acids, the two most important prerequisites for most shift reagent purposes. We have demonstrated (3,8) that the fluorinated chelates such as those of fod are much more effective Lewis acids than unfluorinated analogs. This has been confirmed unequivocally both by gas chromatographic experiments and by NMR (3,7,8).

NMR STUDIES

H(fod) Spectra

NMR spectra (^{19}F and ^{1}H) of Eu(fod)$_3$ and H(fod) in CCl$_4$ solutions are shown in Figure 1. Integration of the H(fod) proton spectrum shows that the compound is essentially 100% enolized under these conditions. Expansion

of the ^{19}F quartet assigned (6) to the CF_2 group next to
the carbonyl reveals a secondary doublet splitting of 0.8
Hz. Since fine structure is also evident in the resonance
of the methine proton, the 0.8 Hz secondary splitting is
probably due to coupling with the methine proton. As
previously reported (6), no detectable spin coupling to
the CF_3 and $CF_2C(O)$ groups was observed for the central
CF_2 group. Very small coupling constants for perfluoro-
alkyl groups may result from mutual cancellation by two
pathways for the transmission of spin information.

Oligomer Formation

In the ^{19}F spectrum of the anhydrous shift reagent,
shown in Figure 2, one of the CF_2 resonances (corresponding to the quartet in the H(fod) spectrum) is not
perceptible above the baseline noise (9). The two
resonances which are apparent (corresponding to the triplet
and singlet) are broad and featureless. Integration of
the spectrum indicates that a very broad CF_2 resonance
(corresponding to the missing "quartet") is also present.
Upon the addition of a Lewis base such as 4-heptanone the
very broad CF_2 resonance becomes sharper and is shifted
upfield. In fact, all of the resonances become sharper
and appear more like the spectrum observed for the free
ligand. Such effects are likely due to the changing
stereochemistry and degree of oligomerization of the
$Eu(fod)_3$ chelate upon adduct formation.

Addition of H(fod) to Solutions of the fod Chelates

$Eu(fod)_3$ is thought to exist in solution as an
oligomer, $(Eu(fod)_3)_n$, where n is probably ≤ 3 (10).
Addition of H(fod) to a solution of $Eu(fod)_3$ might be
expected to cause formation of the seven- or eight-
coordinate adduct $H(Eu(fod)_4)$ causing dissociation of the
$(Eu(fod)_3)_n$ oligomer. We have observed that there is a
rapid exchange on the NMR time scale of fod ligands
between $Eu(fod)_3$ and H(fod), indicating that a tetrakis
complex may exist. The evidence for this exchange interaction is the observation that only one t-butyl resonance
exists in the proton spectra of deuterochloroform
solutions containing both $Eu(fod)_3$ and H(fod). The

chemical shift of this resonance depends upon the relative amounts of the two fod-containing species, showing that this single resonance is not caused by an accidental chemical shift equivalence.

The addition of H(fod) to deuterochloroform solutions of $Eu(fod)_3$ produces small changes in the shifting ability of the latter. Figure 3 shows the effect on the induced shift, $\Delta\delta$, of incremental addition of H(fod) to a solution containing $Eu(fod)_3$/ethanol. For the methylene quartet, where the largest effect is observed, the decrease in shifting power amounts to less than 5% of the total shift. Two plausible explanations for this behavior are apparent: 1. H(fod) as a base cannot compete with ethanol effectively and is displaced, or 2. an $H(Eu(fod)_4)$ species is generated and this can function as a shift reagent by forming a species with coordination number 8 or 9. It has been previously demonstrated that a lanthanide ion, Ce(IV), can accomodate four bulky β-diketonate ligands to form a rather stable compound, $Ce(thd)_4$ (11). A coordination number of 9 has also been observed for lanthanide compounds (12) as illustrated by the crystal structure of $Nd(OH_2)_9(BrO_3)_3$ (13). Therefore the $H(Eu(fod)_4)$ complex may exist as an eight-coordinate species with all of the oxygens coordinated or a seven-coordinate species with one of the oxygens protonated and function as a shift reagent by adding an eighth or ninth donor atom from the substrate.

In order to further investigate the exchange reactions occurring in $Ln(fod)_3$ solutions, approximately equimolar amounts of $Eu(fod)_3$ and $Pr(fod)_3$ were dissolved in carbon tetrachloride, and the NMR spectra (^1H and ^{19}F) were obtained. The ^1H spectra are shown in Figure 4. It can be seen that the t-butyl resonance is very broad and occurs at a chemical shift position intermediate to that for the respective chelates:

$$Eu(fod)_3 + Pr(fod)_3: \quad \delta = 1.27 \text{ ppm, } W = 34 \text{ Hz}$$

where W is the width at one-half maximum. In contrast, the resonances of the t-butyl groups of $Eu(fod)_3$ and $Pr(fod)_3$ separately in CCl_4 are relatively sharp:

$$Pr(fod)_3: \quad \delta = 0.40 \text{ ppm, } W = 8 \text{ Hz}$$

$$Eu(fod)_3: \quad \delta = 1.71 \text{ ppm, } W = 3.5 \text{ Hz}$$

24

These data show a rapid exchange of fod ligands between
$Pr(fod)_3$ and $Eu(fod)_3$, as well as implying an fod exchange
from one Eu(III) to another and from one Pr(III) to
another.

Attempts to resolve the broad t-butyl peak of the
mixture by lowering the temperature to $-20^{\circ}C$ were unsuc-
cessful due to the interaction between the electron and
nuclear magnetic moments, which increases as the
temperature is decreased. The resulting increased paramagnetic broadening more than cancels the effect of the
less rapid exchange rate at lower temperatures. Analysis
of the corresponding ^{19}F spectra leads to the same
conclusions. Figure 5 shows further evidence for the
rapid exchange of fod ligands between $Pr(fod)_3$ and $Eu(fod)_3$.
The ^{19}F spectra shown are of a $Pr(fod)_3$ - $Eu(fod)_3$ mixture
in CCl_4 solution before and in the course of the addition
of a few drops of tetrahydrofuran. Addition of that
substrate causes the exchange-averaged spectrum to be
resolved into the characteristic resonances for adducted
$Eu(fod)_3$ and $Pr(fod)_3$. It is likely that the rapid inter-
complex exchange of fod ligands occurs by means of dimer
or oligomer association.

Effectiveness of Anhydrous Vs. Hydrated Shift Reagent

As ordinarily prepared, $Eu(fod)_3$ is obtained as the
monohydrate, $Eu(fod)_3(H_2O)$. We expected that the
anhydrous chelate might be a more effective shift reagent
because there is no extra donor (i.e., water) present to
compete with the substrate for coordination sites on the
europium. On the other hand it may not be imperative that
the water be displaced in order for other substrates to
become bonded because octacoordinate species are well
known in lanthanide chemistry. Because it is sometimes
inconvenient to work with rigorously anhydrous reagents,
some tests were made to see whether the hydrate will also
function acceptably as a shift reagent. For reference, a
sample of $Eu(fod)_3(H_2O)$ was dried over P_4O_{10} in vacuo and
tested with various substrates. Some of the data are
shown graphically in Figure 6. This figure shows a plot
of the induced shift vs. the shift reagent/substrate mole
ratio, E/S, for the substrate pinacolone, 3,3-dimethyl-2-
butanone. From the slopes of the plots for the monohydrate
and anhydrous shift reagents, it can be argued that the

25

latter is about twice as effective as the former in shifting ability. In similar comparisons it was found that the
$Eu(fod)_3$ which had been dried over P_4O_{10} was about 1.6 times
as effective for shifting the spectrum of 4-heptanone, 1.5
times as effective for dimethyl sulfoxide, and 2.7 times as
effective for acetone as the $Eu(fod)_3(H_2O)$.

There are at least two possibilities that can be
advanced to account for the ability of $Eu(fod)_3(H_2O)$ to
function as a shift reagent.

$$Eu(fod)_3(H_2O) \; + \; S \; \leftrightarrows \; Eu(fod)_3(S) \; + \; H_2O \quad \text{displacement}$$

$$Eu(fod)_3(H_2O) \; + \; S \; \leftrightarrows \; Eu(fod)_3(H_2O)(S) \qquad \text{addition}$$

It is concluded that anhydrous $Eu(fod)_3$ produces larger
shifts, but is more difficult to prepare, keep, and use
than $Eu(fod)_3(H_2O)$. Consequently for non-critical spectral
clarification purposes, from the standpoint of time and
effort it may be quite acceptable to add a little more of
the shift reagent rather than take elaborate precautions
to dry and keep rigorously dry the shift reagent, solutions,
NMR tube, etc.

Stability of $Eu(fod)_3$ with Carboxylic Acids

It has been reported (14) that $Eu(thd)_3$ is unstable in
solution with weak acids. In contrast, we have found that
moderate concentrations of $Eu(fod)_3$ are stable in solution
with a carboxylic acid. We have been able to produce shifts
with $Eu(fod)_3$ which are unchanged over a seven-day period
in the spectra of n-butyric acid dissolved in CCl_4. (In
these experiments n-butyric acid was selected only for
convenience; it gives a reasonably well-defined spectrum
in carbon tetrachloride solutions without any shift
reagent.) The stability of dilute $Eu(fod)_3$ solutions to
n-butyric acid and the apparent ability of the carboxyl
group to coordinate with $Eu(fod)_3$ suggest that the spectra
of carboxylic acids can be simplified using moderate
concentrations of $Eu(fod)_3$ with negligible decomposition of
either the acid or the shift reagent.

In another experiment the proton spectrum of N-trifluoro-
acetyl-d-alanine in CCl_4 was altered by adding $Eu(fod)_3$.
After the $Eu(fod)_3$ was added the spectrum remained unchanged over a several-hour period, indicating that this

shift reagent is not readily decomposed by amino acid derivatives. This suggests that the fod chelates may be useful in studies of peptides and derivatized peptides with blocking groups. C-end group analysis and peptide sequencing may ultimately prove feasible.

Comparison of Eu(fod)$_3$ with Eu(thd)$_3$

A comparison of the efficiencies of Eu(fod)$_3$ and Eu(thd)$_3$ in separating the resonances of the methoxy protons of an isomeric mixture of unsymmetrically para-substituted azoxybenzenes was made by adding approximately equimolar amounts of the two shift reagents to identical solutions of the isomeric mixture. Figure 7 shows the spectra of the methoxyl and benzylic methylene resonances of a solution of the isomeric 4(and 4')-n-butyl-4'(and 4)-methoxy-azoxybenzenes before and after the addition of each of the two shift reagents (3). After addition of only 10 mg of Eu(fod)$_3$ the resonances for the two isomers are easily identified. In contrast, addition of 10 mg of Eu(thd)$_3$ does not induce any observable shifts under the same conditions. Presumably the ineffectiveness of Eu(thd)$_3$ in this case is due to its poorer Lewis acidity relative to that of Eu(fod)$_3$.

Interestingly, the methoxy proton resonances are shifted in opposite directions for the two isomers. In one of the isomers the expected downfield shift is induced by complex formation with Eu(fod)$_3$. With the other isomer the shift of the methoxy proton resonances is upfield; this is presumably due to the greater angle θ_i in the McConnell-Robertson equation (vide infra). At angles greater than 55° but less than 125° the $3 \cos^2 \theta_i - 1$ term becomes negative and upfield shifts result.

Other New Shift Reagents

While the chelates of fod ordinarily function quite effectively as Lewis acids, there may be instances in which even more acidic shift reagents are desired for use with very weak nucleophiles. By maintaining a perfluoro-propyl group at one end of the ligand and substituting fluorines for methyl groups, Richardson and Sievers (15) have created the highly acidic and soluble lanthanide chelates of 1,1,1,5,5,6,6,7,7,7-decafluoro-2,4-heptanedione

$$F_3C-C-\overset{\ominus}{\underset{\underset{H}{|}}{C}}-C-CF_2-CF_2-CF_3 \qquad \text{dfhd}$$

With N,N-dimethylaniline the dfhd chelate of Eu(III) induces greater shifts than the fod or thd chelates on an equimolar basis. The solubility of the dfhd chelates is phenomenal; in one experiment 260 mg of the Eu(III) complex was added to a solution containing 20 mg of n-heptanol in 0.5 ml CCl_4. There was no appreciable degradation in resolution in spite of the enormous amount of the chelate present, and a first order spectrum was easily generated.

On occasion the large t-butyl resonances from the fod and thd chelates appear in a region of the spectrum that interferes with the resonances of the substrate. This can be overcome by using dfhd or by employing deuterated fod chelates.

GAS CHROMATOGRAPHIC STUDIES

Gas chromatography offers an independent method of examining the interactions between nucleophiles and lanthanide shift reagents. The success of this method depends upon the ability to incorporate the lanthanide shift reagent into the stationary phase of a GC column. Solubility in organic solvents is an important factor in determining the applicability of a compound to studies of this nature. Fortunately many β-diketonate chelates possess this property as well as thermal stability at normal column operating temperatures. The technique involves dissolving the metal chelate in a suitable organic liquid phase (e.g., squalane) and coating a support material with this solution. Organic nucleophiles are then passed through a column packed with this coated support, and retention times are corrected for dead volume and normalized using the retention time of some solute which does not interact with the metal chelate (e.g., methyl-cyclohexane). For comparison purposes retention times are

obtained for each of the nucleophiles on a column coated
with the liquid phase minus the metal chelate.

Theory

The theory for the study of interactions of this nature
has been developed with the assumption that the following
equilibrium exists in the liquid phase of the GC column
(8,16):

$$S(solution) + chelate \overset{K}{\leftrightarrows} chelate \cdot S$$

$$K = \frac{[chelate \cdot S]}{[S(solution)][chelate]}$$

where S(solution) represents donor molecules in solution
but not complexed with the chelate, and chelate·S
represents the adduct formed between the chelate and the
donor. If extremely small donor sample injections are
made (e.g., headspace vapors), the concentration of
chelate in solution is in such great excess that it is
essentially constant. Then one can obtain a pseudo
equilibrium constant K' which is related to K by the
expression:

$$K' = K[chelate] = \frac{[chelate \cdot S]}{[S(solution)]} = \frac{[complexed]}{[free]}$$

Therefore, the physical significance of K' is that it provides a means for determining the fraction of time that a
donor molecule is complexed while in solution.

As stated earlier this derivation is based on the
equilibrium shown above (i.e., a 1:1 complex between
donor and chelate is assumed). In the experiment the
excess of chelate with respect to donor tends to favor
1:1 adduct formation, but the possibilities that other
species may be formed or that chelate molecules may
exist as more highly associated species are not totally
discounted. If other equilibria are involved the relationship between K and K' is altered. Therefore, one should
not assume the validity of K values without prior assur-
ance that no other equilibria are involved. Despite these
apparent limitations, the values of K' give valuable information about the relative stabilities of the adducts

29

formed. It is in this respect that useful information can be obtained concerning the interactions between lanthanide shift reagents and nucleophiles.

It has been shown (8,16) that K' is directly related to retention data by the following expression:

$$K' = (t_{chelate}/t_{squalane})-1$$

where $t_{chelate}$ is the corrected normalized retention time of a given nucleophile on a column using a solution of the chelate in squalane as the liquid phase, and $t_{squalane}$ is the corrected normalized retention time of that nucleophile on a column using squalane as the liquid phase. Extensive GC studies have been conducted to examine the interactions of selected ketones and alcohols with the NMR shift reagent, $Eu(fod)_3$ (17).

In general alcohols form more stable complexes than ketones with $Eu(fod)_3$. This is indicated by the generally higher K' values obtained for alcohols (e.g., the K' values for 2-butanone and 2-butanol and 14.6 and 20.8 respectively). Several trends appear in the data which may be explained on the basis of relative nucleophilicities and steric constraints of the interactions. These trends have been summarized in terms of four effects: chain lengthening, functional group position, branching, and cyclization.

Chain Length

The data indicate that, as a general rule, increasing the length of the alkyl chain in a series of similar compounds increases the stability of the complex formed. For example, the K' values for 2-propanol, 2-butanol, 2-pentanol, 2-hexanol, and 2-heptanol are 20.8, 20.8, 22.1, 22.7, and 23.2 respectively. These results are consistent with the expected increase in nucleophilicity due to the increased inductive effect of the lengthened alkyl chain. There are indications that this effect reaches a limiting value beyond which the added inductive effect becomes so small that steric constraints become predominant. In the analogous 2-ketone series the K' value increases up through 2-hexanone and then falls off slightly as the alkyl chain is further lengthened.

Another important factor which determines the strength of the interaction is the position of the functional group within the molecule. This effect can be attributed to steric constraints upon the interaction. The K' values indicate that the most stable complexes are formed with donor molecules in which the functional group involved in the interaction is close to one end of the molecule. The stabilities of the complexes decrease as the functional group is moved closer to the center of the molecule (e.g., the K' values for 2-heptanone, 3-heptanone, and 4-heptanone are 16.4, 14.2 and 13.5 respectively).

Branching

Branching in a donor molecule is a factor which is more difficult to analyze, because it involves opposing influences. On the one hand a molecule becomes a better donor due to the increased inductive effect of the branching, while, at the same time, the interaction becomes less sterically favorable. Comparison of K' values for ketone isomers which vary only in the degree and/or position of branching show that, in general, branching with methyl groups at the carbon atoms α or γ to the keto group results in more stable complexes than the non-branched isomer, and branching with a methyl group at the β carbon atom results in a less stable complex. This indicates that inductive influences predominate in the interactions of ketones that are branched at the carbon atoms α or γ to the keto group and steric constraints predominate for β branching. The interactions of alcohols are more critically affected by steric influences than are ketones. This is illustrated by the fact that branching at any position results in less stable complexes for branched alcohols compared to their non-branched isomers. This is reasonable when one considers that the oxygen atom in an alcohol is bonded to a hydrogen atom, creating additional steric constraints on the interactions which are not present in ketones. In fact, even though alcohols generally form more stable complexes than ketones with $Eu(fod)_3$, in highly branched systems the situation can be reversed (e.g., the K' values for 3,3-dimethyl-2-butanone and 3,3-dimethyl-2-butanol are 18.7 and 16.1 respectively).

Cyclization

Cyclization is another way of increasing the stability of the $Eu(fod)_3$-substrate complex. Apparently, tying back the ends of the molecule creates a more sterically favorable approach for the Eu-O interaction. Thus the K' value for cyclohexanone is 23.2 compared to 16.9 for 2-hexanone. By examining data of this type one can predict the preferred site of bonding in a multi-functional substrate. For example one would expect preferred bonding in a steroid derivative to occur through an unhindered ring keto group rather than through a linear side chain keto group.

Increased Affinity of Substrates for Fluorinated Shift Reagents

Other more extensive gas chromatographic studies of various lanthanide chelate-donor interactions (8) have revealed several important results which are pertinent to NMR shift studies. These studies showed that Er(III) chelates of fluorinated β-diketones (e.g. fod and facam) undergo much stronger interactions with nucleophiles than similar nonfluorinated chelates. This is most likely due to increased complex acidity in the fluorinated chelates which is in turn due to a decrease in the ligand basicity. Thus, it could be predicted from GC data that $Eu(fod)_3$ would form stronger complexes and be a superior shift reagent for weak nucleophiles compared with $Eu(thd)_3$, which is clearly a poorer Lewis acid.

Trends in the Lanthanide Series

An interesting correlation was obtained for the effect of size of the chelate metal ion on the interaction. In a series of $Ln(facam)_3$ chelates it was found that for strong donors, such as THF, the retention times increased exponentially with the inverse of the ionic radius of the metal. This has been interpreted as resulting from a change in the extent of aggregation of the chelates with the reciprocal of the ionic radius. Further evidence supporting this is given by retention data for the hydrolyzed chelates $Er(OH)(dibm)_2$ and $Er(OH)(acam)$ which indicate that these chelates form weaker complexes with nucleophiles than does $Er(thd)_3$. Molecular weight data indicate that these hydrolyzed chelates are strongly associated and are dimeric at concentrations at which monomeric $Er(thd)_3$ predominates. The lanthanide contraction may also cause more stable complexes to be formed by virtue of the shorter bond lengths of the heavier lanthanides.

32

Isotope Effects

Gas chromatographic data have also provided independent confirmation of the isotope effect observed by Hinckley and co-workers in NMR shift experiments (18,19). Alcohols which are deuterated geminally to the hydroxyl group have longer retention times than their non-deuterated isomers on columns using $Eu(fod)_3$ in squalane as the liquid phase (17). Identical columns without $Eu(fod)_3$ gave virtually identical retention times for both the deuterated and non-deuterated isomers. In the case of 2-butanol and d_1, 2-butanol, a difference of 7.1 sec or 3.5% was observed. These differences are the result of differing $Eu(fod)_3$-alcohol interactions. Hinckley has postulated two possible explanations for this phenomenon: (1) the existence of a hydrogen bond between the geminal proton (or deuteron) and an oxygen atom of one of the β-diketonate groups of the metal chelate, or (2) increased basicity of the hydroxyl oxygen atom due to geminal deuteration. Data has been published which indicate that the effective inductive electron release from CD_3(or CD_2) is greater than from CH_3(or CH_2) (20). These results give support to the second of Hinckley's explanations. Our studies do not provide a means for distinguishing between the two possibilities. However, the decrease in stability of complexes of tertiary alcohols with $Eu(fod)_3$ compared to those with secondary alcohols could possibly be in part due to the fact that tertiary alcohols have no geminal proton available to form the hydrogen bond postulated in the first of Hinckley's explanations.

X-RAY STRUCTURAL STUDIES

Application of the modified McConnell-Robertson relationship (21)

$$\Delta H_i/H = K (3 \cos^2 \theta_i - 1)/r_i^3$$

for calculating relative induced shifts requires a knowledge of the detailed molecular stereochemistry of the substrate/shift reagent complex. In particular, the required parameters are the distance from the resonating nucleus to the paramagnetic ion, r_i, and the angle, θ_i, which this vector makes with the principal magnetic axis of the molecule. For this simplified relationship to be

33

applicable the magnetic moment of the complex must have axial symmetry (i.e., a C_3 or higher axis of symmetry). Structural determinations by x-ray diffraction techniques offer a means for obtaining the necessary parameters of the complex in the crystalline state; we have, therefore, undertaken a systematic study of the crystal structures of selected mono-and bis-adducts of Eu(thd)$_3$. In addition to the parameters required for calculating induced shifts, basic underlying structural principles may also be deduced from the resulting crystal structures. Several of the more important data which are available from structure determinations are: ranges of bond lengths for the Eu-O (β-diketone) and Eu-X (adduct) bonds, bond lengths and angles within the adduct molecules, and the overall stereochemistry of the shift reagent complex. These data may be of invaluable aid in the determination of preferred conformations in solutions by allowing the experimenter to formulate more realistic a priori assumptions concerning the configurations of the species in solution.

X-ray Measurements

Crystals of Eu(thd)$_3$(DMF)$_2$ and Eu(thd)$_3$(DMSO) suitable for diffraction studies were obtained from the slow evaporation of saturated solutions of Eu(thd)$_3$ dissolved in the appropriate neat solvent. Both compounds are unstable and lose the adducted molecules DMF and DMSO upon prolonged exposure to the atmosphere.

Preliminary precession photographs of single crystals sealed in thin-wall glass capillaries established the probable space groups, orientations, and unit cell parameters for each of the compounds. Despite the fact that mother liquor was sealed in the capillaries to retard decomposition, it was necessary to use five separate crystals of Eu(thd)$_3$(DMF)$_2$ during the course of data collection; two crystals were required for the study of Eu(thd)$_3$(DMSO). Intensity data were collected with a computer-controlled Picker FACS-1 diffractometer using graphite monochromatized MoK$_\alpha$ radiation and the θ-2θ scan technique; background counts were measured at the scan limits for each reflection. After being corrected for background and incident beam polarization effects, the data for each compound were directly reduced to relative scattering amplitudes, $|Fo|$. Corrections for absorption effects were applied only to the data for Eu(thd)$_3$(DMSO).

34

Both structures were solved by the usual heavy-atom methods and refined using the block-diagonal approximation to the method of least squares. The choice of space group for each compound is supported in every detail by the successful refinement of the structure. One hundred independent carbon and heavier atoms were refined for the structure of $Eu(thd)_3(DMF)_2$ and eighty-eight atoms were refined in the structure of $Eu(thd)_3(DMSO)$.

Structures of $Eu(thd)_3(DMF)_2$ Isomers

A centrosymmetric triclinic unit cell of $P\bar{1}$ symmetry contains four monomeric $Eu(thd)_3(DMF)_2$ molecules which are grouped in two non-equivalent pairs; within each pair the molecules are related by a center of symmetry. The Eu(III) ion is bonded to the eight oxygen atoms of three bidentate β-diketonate ions and the two monodentate DMF molecules. In contrast to the coordination geometry displayed by the bis adducts of 4-picoline (22) and pyridine (23), the DMF ligands are cis on the same square face of a distorted square-antiprism. The thd moieties are bound such that no edge joining the square faces of the coordination polyhedra is spanned. A similar molecular stereochemistry is exhibited by both crystalline $La(acac)_3(H_2O)_2$ (24) and $[Y(acac)_3(H_2O)_2]\cdot H_2O$ (25).

Structures of $Eu(thd)_3$ DMSO Isomers

Four discrete hepta-coordinate molecules of $Eu(thd)_3(DMSO)$ (two geometrically non-equivalent pairs, vide infra) crystallize in a triclinic unit cell of $P\bar{1}$ symmetry. Each β-diketonate ion is bonded to the Eu(III) ion as a bidentate ligand and the DMSO is also coordinated through an oxygen atom.

Differences among the various idealized seven coordinate geometries are minor and a distorted polyhedron can often be equally well described in terms of several of these (12,26,27). The observed coordination polyhedra of the DMSO complex can best be visualized in terms of either distorted pentagonal bipyramids or trigonal base-tetragonal base geometries. The latter description is directly derivable from the distorted square-antiprisms displayed in $Eu(thd)_3(DMF)_2$ by the removal of one unidentate ligand and positioning the remaining one mid-way along the edge. Minor repositioning of the bidentate ligands are also required to achieve the observed stereochemistry. If the two isomers of the DMSO adduct are considered as distorted pentagonal bipyramids by ignoring the chelate rings and the rufflings of the equatorial planes, each complex has what approximates a five-fold axis colinear with the bond linking the DMSO oxygen and the europium atoms. The pentagonal bipyramidal structure is depicted in Figure 7 with the chelate rings omitted for clarity. It must be emphasized that while axial symmetry was not found in the solid state it is very easy to visualize a time-averaged five-fold rotation axis in solution.

A common phenomenon observed in metal/β-diketonate structures is ring folding; the folding of the ring is along the line joining the two oxygen atoms. In both structures studied, folding is prominent, and the angle of bending is different for each of the twelve metal chelate rings. The observed range of folding is from 2° to 21° and apparently is primarily determined by crystal packing effects.

In both complexes the t-butyl groups of the thd ligands are characterized by large thermal motion and several orientations of the methyl groups with respect to the ring backbones are observed. The conclusion drawn is that in the unconstrained molecules there is virtually free rotation about the σ bonds joining the t-butyl groups to the ring backbones.

Implications of the X-ray Data on the Question of Effective Axial Symmetry in Solution

In both crystal structures studied, chemically equivalent molecules have been rendered geometrically non-equivalent by crystal packing effects. For each complex two pairs of geometrically distinct molecules are contained in the same unit cell in spite of the fact that the europium ions are surrounded by the same donors in the same ratios. Both the coordination polyhedra and the chelate ring orientations show significant differences. Dimensions of the non-equivalent coordination polyhedra for both Eu(thd)$_3$ (DMF)$_2$ and Eu(thd)$_3$(DMSO) are schematically diagramed in Figure 8 and Figure 9, respectively. The differences in analogous dimensions of the polyhedra clearly indicate the ease with which these higher coordinate complexes can be deformed. In particular, the seven-coordinate complex (presumably the most probable species in solution (28)) shows

36

the greatest differences between supposedly similar edges
and angles of the nonequivalent polyhedra. Releasing
the molecules from the constraints imposed by the
crystal lattice presumably allows the geometrically non-
equivalent polyhedra to assume a single, time-averaged
configuration not necessarily related in all detail to
either of the two solid state configurations.

Correlation of the crystal structures with the
solution conformation is further complicated by facile
ligand exchange (both the adduct and β-diketonate ligands
undergo fast exchange on the NMR time scale) and intramolecular reorganizations. Low potential energy barriers
between the idealized higher coordinate polyhedra might
also permit the time-averaged solution configuration to
be significantly different from that displayed in the
solid state.

Only two complexes of lanthanide shift reagents and
related compounds have been shown to possess or approximate
quite closely highly symmetric coordination polyhedra;
the monocapped octahedron of $Ho(dbm)_3(H_2O)$ (29) is rigorously required (crystallographically) to have three-fold
symmetry in the solid state, and the coordination geometry
of $Er(thd)_3$ (30), as shown by x-ray diffraction, is almost
ideally trigonal prismatic. The persistence of this
symmetry in solution is open to some question in view of
the ease that complexes of this type can be deformed.
Conversely, because rearrangement of seven-and eight-
coordinate lanthanide complexes is so facile, the absence
of three-fold symmetry may often be due to crystal packing
distortions, and the complex may actually have a higher
symmetry in solution than in the solid state.

The octacoordinate anion $Mo(CN)_8^{3-}$ offers a striking
example of the effects of crystal environments upon the
observed geometries. In the solid state the arrangement
of the eight cyano groups about the Mo(V) ion is dependent
upon the anion present: in the $(n-butyl)_4N^+$ salt, the
geometry is rigorously D_2 but closely approximates D_{2d} (31);
a polyhedron approximating D_{4d} symmetry has been observed
in the salt $Na_3W(CN)_8 \cdot 4H_2O$ which is isomorphous with the
Mo(V) complex (32).

Blight and Kepert (33) stress that factors such as
crystal field stabilization, ligand-ligand repulsion, and
solvation energies could not be successfully correlated to
the preferred stereochemistry of higher coordinate complexes and that lattice energy of the crystal appears to be

37

the most important factor in ultimately deciding the conformation exhibited in the solid state. Consequently, the
fact that the solid state structure is not rigorously
axially symmetric need not necessarily mean that the
simplified McConnell-Robertson equation will not hold in
solution.

Metal-Substrate Bond Lengths

The available x-ray structural data clearly show the
Ln-O(β-dik) bond lengths lie within the expected ranges.
For Eu(III), in particular, this range is approximately
2.30-2.40Å with 2.35Å as the average. In general a valid
first approximation for this bond length can be obtained
from the sum of the Pauling empirical crystal radius for
the Ln(III) ion and the radius of the oxide ion, 1.34Å. The
Ln-X(adduct) bond lengths are naturally dependent not only
upon the radius of the lanthanide ion but also upon the
nature of the coordinating atom, X.

Bond lengths of the type Ln-X where X represents either
an oxygen or a nitrogen atom have been compiled from many
crystal structure reports by Sievers, et al. (34) for a broad
variety of lanthanide compounds and several pertinent conclusions may be drawn from these data. It is apparent that
the values of these bond lengths fall within rather narrow
ranges considering the variety of complexes and coordination
numbers displayed by the compounds included in the compilation. For all of the lanthanide (III) complexes considered,
median values for the Ln-O and Ln-N bonds are 2.45 and 2.60Å
respectively; a range of ±0.3Å about each of these values
encompasses virtually all reported Ln-O and Ln-N bond
distances. A clear trend with the heavier atoms having
shorter bond distances, mirroring the lanthanide contraction,
can be detected when all of the x-ray data are examined,
but in essentially all cases the bond lengths from the
metal to nitrogen and oxygen donors lie between 2.1 and 2.9Å.
In view of these data any treatment of lanthanide induced
shift data should be based on the a priori assumption that
the lanthanide-substrate bond length lies within the cited
ranges.

CONCLUSIONS

Several important results have been obtained from the studies we have conducted. The variations in coordination geometries for chemically equivalent molecules of shift reagent-substrate complexes in the solid state illustrate the ease with which changes occur in the coordination sphere of lanthanide complexes. It is reasonable to conclude that these variations occur even more readily in solution, leading to the possibility that effective, time-averaged axial symmetry can be attained in solution by lanthanide complexes which do not possess rigorous axial symmetry in the solid state. This seems to be the most plausible explanation for the success of the simplified McConnell-Robertson relationship in predicting molecular configurations in solution.

Gas chromatographic studies have provided independent confirmation of the isotope effect which Hinckley observed in NMR studies. They have also shown that lanthanide chelates of fluorinated β-diketonate ligands form more stable complexes with nucleophiles than do chelates of non-fluorinated ligands. Thus, GC data were used effectively to confirm the superior shift reagent qualities of $Eu(fod)_3$ compared to $Eu(thd)_3$.

Although $Eu(fod)_3(H_2O)$ is effective as a shift reagent, NMR data have shown that $Eu(fod)_3$ which has been dried in vacuo over P_4O_{10} is superior in shifting ability. Nevertheless, the relative ease of preparing, storing, and handling $Eu(fod)_3(H_2O)$ might make its use acceptable for all but the most critical shift reagent applications.

In view of these findings, it is appropriate to include a brief discussion of the principles of selection of shift reagents (3). $Eu(fod)_3$ is probably the best overall shift reagent for spectral clarification presently available. Its ability to form more stable complexes than $Eu(thd)_3$ with nucleophiles makes it more useful for weak donors. This, coupled with its superior solubility (about 10 times as soluble as $Eu(thd)_3$ in common NMR solvents), makes it the most likely candidate for general applications. Other chelates produce greater shifts (e.g., $Ho(fod)_3$ which shifts upfield) but produce considerably broader peaks. Thus, for some applications it may be advantageous to use $Ho(fod)_3$ if one is not interested in fine structure. For extremely weak nucleophiles one can take advantage of the fact that nucleophiles form more

stable complexes with lanthanide chelates as the ionic
radius of the metal decreases. This means that $Yb(fod)_3$
is expected to form more stable complexes than $Eu(fod)_3$
and could possibly be used more effectively for very weak
donors. Since $Yb(fod)_3$ has been reported to be essentially
monomeric in solution (35), and because contact contributions are small for Yb(III) this complex may offer special
advantages in structural studies. As a general rule,
lanthanide shift reagents that produce downfield shifts
have been most widely used. If it is desirable to obtain
an upfield shift, $Pr(fod)_3$ is probably the best shift
reagent available. Special applications have resulted
in the use of shift reagents which possess special
characteristics. For example, lanthanide chelates of
optically active camphor derivatives have been used to
produce different shifts in racemic solutions of optically
active compounds (36), therby allowing measurement of
optical purity by NMR.

ACKNOWLEDGEMENT

We gratefully acknowledge the contributions of
Dr. Larry R. Froebe to these research efforts.

REFERENCES

1. C.C. Hinckley, J. Amer. Chem. Soc., 91, 5160 (1969).
2. M.R. Wilcott, III, R.E. Lenkinski, and R.E. Davis,
   J. Amer. Chem. Soc., 94, 1742 (1972).
3. R.E. Rondeau and R.E. Sievers, Analytical Chemistry,
   in press.
4. Abbreviations used in this manuscript are: fod, the
   anion of 6,6,7,7,8,8,8-heptafluoro-2,2-dimethyl-3,5-
   octanedione; thd, the anion of 2,2,6,6-tetramethyl-
   3,5-heptanedione; facam, the anion of 3-trifluoro-
   acetyl-d-camphor; dfhd, the anion of 1,1,1,5,5,6,6-
   7,7,7-decafluoro-2,4-heptanedione; dibm, the anion of
   2,6-dimethyl-3,5-heptanedione; acam, the anion of
   3-acetyl-d-camphor; dbm, the anion of 1,3-diphenyl-
   propanedione; acac, the anion of 2,4-pentanedione;
   DMF, N,N-dimethylformamide; DMSO, dimethylsulfoxide.
5. K.J. Eisentraut and R.E. Sievers, J. Amer. Chem. Soc.,
   87, 5254 (1965).
6. C.S. Springer, Jr., D.W. Meek and R.E. Sievers,

Inorg. Chem., 6, 1105 (1967).

7. B. Feibush, R.E. Sievers, and C.S. Springer, Jr., Abstracts, 158th National A.C.S. Meeting, New York, Sept. 1969.

8. B. Feibush, M.F. Richardson, R.E. Sievers, C.S. Springer, Jr., J. Amer. Chem. Soc., 94, 6717 (1972).

9. L.R. Froebe, J.J. Brooks, R.E. Sievers, and D.S. Dyer, Proc. XIV Int. Conf. on Coord. Chem., Toronto, June 1972.

10. J.F. Desreux, L.E. Fox and C.N. Reilley, Analytical Chem., 44, 2217 (1972); C.S. Springer, Jr., A.H. Bruder, S.R Tanny, M. Pickering, and H.A. Rockefeller, Chapter 14.

11. J. Selbin, N. Ahmad and N. Bhacca, Inorg. Chem., 10, 1383 (1971).

12. E.L. Muetterties and C.M. Wright, Quart. Rev., (London), 21, 109 (1967).

13. L. Helmholz, J. Amer. Chem. Soc., 61, 1544 (1939).

14. J.K.M. Sanders and D.H Williams, J. Chem. Soc., (D), 422 (1970).

15. M.F. Richardson and R.E. Sievers, Inorg. Chem., 10, 498 (1971).

16. M.A. Muhs and F.T. Weiss, J. Amer. Chem. Soc., 84, 4697 (1962).

17. J.J. Brooks and R.E. Sievers, J. Chromatog. Sci., in press.

18. G.V. Smith, W.A. Boyd, and C.C. Hinckley, J. Amer. Chem. Soc., 93, 6319 (1971).

19. C.C. Hinckley, W.A. Boyd, and G.V. Smith, Tet. Let., 10, 879 (1972).

20. E.A. Halevi, M. Nussim, and A. Ron, J. Chem. Soc., 866 (1963).

21. H.M. McConnell and R.E. Robertson, J. Chem. Phys., 29, 1361 (1958).

22. W.DeW. Horrocks, Jr., J.P. Sipe, and J.R. Luber, J. Amer. Chem. Soc., 93, 5258 (1971).

23. R.E. Cramer, and K. Seff, J.C.S. Chem. Comm., 1972, 400 (1972).

24. T. Phillips, II, D.E. Sands, and W.F. Wagner, Inorg. Chem., 7, 2285 (1968).

25. J.A. Cunningham, D.E. Sands, and W.F. Wagner, Inorg. Chem., 6, 499 (1967).

26. T.A. Claxton, and G.C. Benson, Can. J. Chem., 44, 157 (1966).

27. R.J. Gillespie and R.S. Nyholm, Quart. Rev. (London),

11, 339 (1957).

28. S.J. Ghotra, F.A. Hart, G.P. Moss, and M.L. Staniforth, J.C.S. Chem. Comm., 1973, 133 (1973).
29. A. Zalkin, D.H. Templeton, and D.G. Karraker, Inorg. Chem., 8, 2680 (1969).
30. J.P.R. de Villiers, and J.C.A. Boeyens, Acta Cryst., Sect. B., 2335 (1971).
31. B.J. Corden, J.A. Cunningham, and R. Eisenberg, Inorg. Chem. 9, 356 (1970).
32. L.D.C. Bok, J.G. Leipoldt, and S.S. Bassen, Acta Cryst., Sect. B, 684 (1970).
33. D.G. Blight and D.L. Kepert, Theoret. Chim. Acta, (Berlin), 11, 51 (1968).
34. R.E. Sievers, C.S. Springer, Jr., and M.F. Richardson, review of lanthanide chelates, to be published.
35. T.J. Marks, R. Porter, J.S. Kristoff, and D.F. Shriver, Chapter 12.
36. See Chapter 4 for a review of chiral shift reagents.

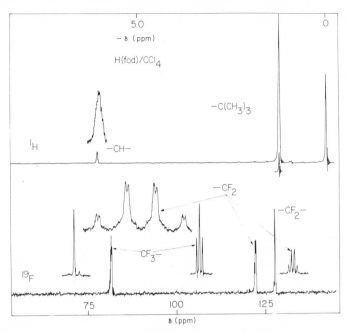

Figure 1. ^{19}F and ^1H spectra (14,092 gauss) of H(fod).

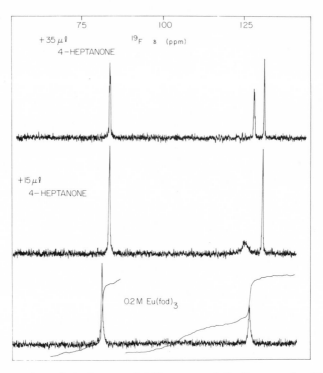

Figure 2. Anhydrous shift reagent before and after addition of 4-heptanone.

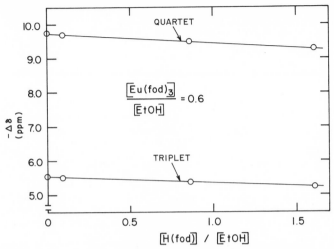

Figure 3. $-\Delta\delta$ <u>vs.</u> $[H(fod)] / [EtOH]$ at E/S = 0.6.

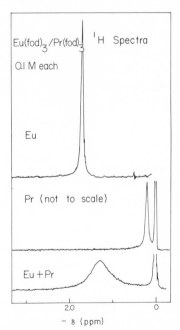

Figure 4. ^1H spectra of $Eu(fod)_3$ (top), $Pr(fod)_3$ (middle), and $Eu(fod)_3$-$Pr(fod)_3$ mixture (below).

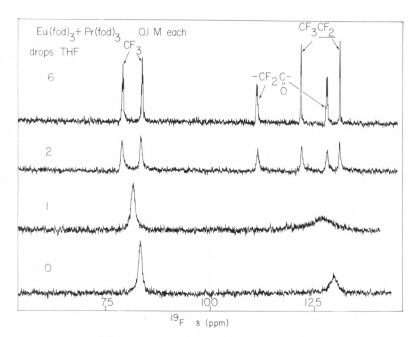

Figure 5. ^{19}F spectra obtained by incremental
addition of tetrahydrofuran to a Pr(fod)$_3$-
Eu(fod)$_3$ mixture.

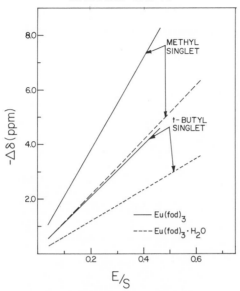

Figure 6. $-\Delta\delta$ vs. E/S for pinacolone with Eu(fod)$_3$ and with Eu(fod)$_3$(H$_2$O).

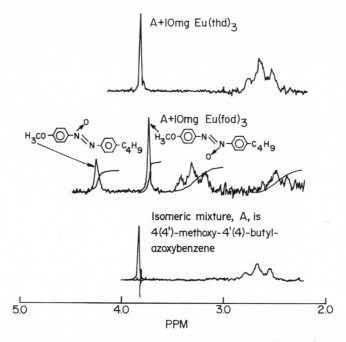

Figure 7. The partial proton spectra (60MHz) of an isomeric mixture in CCl₄ (0.2M) before and after the addition of 10 mg of Eu(thd)₃ or Eu(fod)₃.

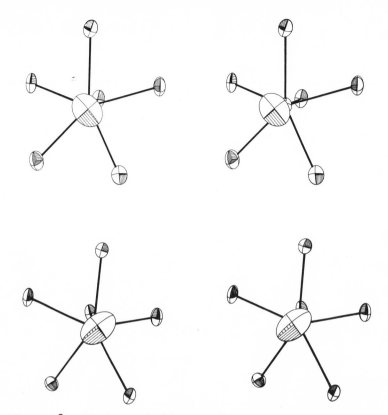

Figure 8.   Pentagonal bipyramid coordination polyhedra
of Eu(thd)$_3$(DMSO) as determined for the two
related but nonidentical isomers by x-ray
structure determination.   The view of the
remainder of the chelate molecule is that
seen by the substrate.   The europium atom
is hidden behind the oxygen atom of the
DMSO moiety.

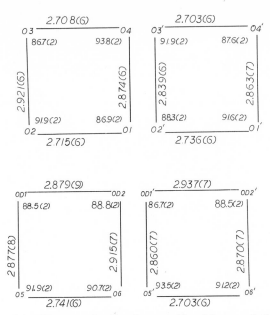

Figure 9. Non-equivalent coordination polyhedra of $Eu(thd)_3(DMF)_2$.

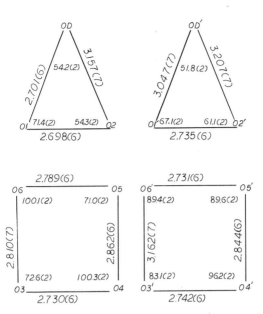

Figure 10. Non-equivalent coordination polyhedra of Eu(thd)$_3$(DMSO).

MAGNETIC ANISOTROPY AND DIPOLAR SHIFTS IN SHIFT REAGENT SYSTEMS

William DeW. Horrocks Jr., James P. Sipe, III and Daniel Sudnick

Department of Chemistry, The Pennsylvania State University, University Park, Penna. 16802

I. INTRODUCTION

The study of the nmr spectra of paramagnetic molecules has been an area of ever increasing activity for over ten years. The principles and major applications of this field are reviewed in a current monograph.[1] Lanthanide shift reagents (LSR's), introduced by Hinckley[2] in 1969, represent the single most important analytical application stemming from research in this area. Aside from articles in this volume, at least two fairly extensive reviews[3,4] are, or shortly will be, available. In addition to effecting simplification and resolution enhancement of the nmr spectra of functional organic substrate molecules, shift reagents have the potential, in principle at least, of yielding valuable information about the geometrical structures and conformations of molecules in fluid solution. The goal of our research in this area has been to obtain a fundamental understanding of the mechanism of action of LSR's with particular attention paid to their potential as probes of structure in solution. Our approach has been both experimental and theoretical. This article outlines our efforts in both of these categories.

Work in these laboratories on the nmr of paramagnetic transition metal complexes has emphasized, to some degree, the separation of observed isotropic shifts into their contact and dipolar contributions.[5-12] One of us has recently pointed out[10] that dipolar shifts may be correctly

evaluated from magnetic susceptibility anisotropy data
rather than from g-tensor anisotropy information as had
been customary previously. This approach has been verified
by independent theoretical treatments.[13,14] We have at-
tacked the problem experimentally by a combination of nmr
and single crystal x-ray and magnetic anisotropy studies.
On the theoretical side we have developed a completely
general method of handling the calculation of paramagnetic
anisotropy in lanthanide complexes and hence of dealing
with dipolar shifts in these systems. In contrast to
alternative approaches,[15] our method may be readily extended to low symmetry situations which appear to be common
in shift reagent systems.

It is appropriate to present in this section the
fundamental equations for the contact, $\Delta H^{con}/H$, and dipolar
$\Delta H^{dip}/H$ contributions to the observed isotropic shifts,
$\Delta H^{iso}/H$, in shift reagent systems, where the last quantity
is, of course, the sum of the first two. The fundamental
equation for the contact shift is

$$\frac{\Delta H^{con}}{H} = \left(\frac{A}{h}\right) \frac{2\pi}{\gamma_N} \langle S_z \rangle \tag{1}$$

where (A/h) in the electron-nuclear hyperfine interaction
constant in Hz, $\langle S_z \rangle$ is the expectation value of the z
component of the total spin, and γ_N is the nuclear magneto-
gyric ratio. For lanthanide systems $\langle S_z \rangle$ may be evaluated
in terms of the total angular momentum quantum number J,
leading to equation 2.

$$\frac{\Delta H^{con}}{H} = -\left(\frac{A}{h}\right) \frac{2\pi}{\gamma_N} \frac{g_J \beta (g_J-1) J(J+1)}{3kT} \tag{2}$$

under a number of not necessarily realistic (in all cases)
assumptions, where g_J is given by equation 3.

$$g_J = 1 + \frac{S(S+1) + J(J+1) - L(L+1)}{2J(J+1)} \tag{3}$$

A generally valid expression for dipolar shifts is
given[11] by equations 4 and 5

$$\frac{\Delta H^{dip}}{H} = -D\left\langle\frac{3\cos^2\theta-1}{r^3}\right\rangle \quad -D'\left\langle\frac{\sin^2\theta\cos2\Omega}{r^3}\right\rangle \tag{4}$$

$$D = \frac{1}{3N}[\chi_z-1/2(\chi_x + \chi_y)]; \quad D' = \frac{1}{2N}[\chi_x - \chi_y] \tag{5}$$

where r, θ, and Ω are the spherical polar coordinates of the resonating nucleus in the coordinate system of the principal magnetic axes, the χ's are the principal molecular susceptibilities, and N is Avogadro's number.

II. SURVEY OF ACTION OF LANTHANIDE SHIFT REAGENTS

Our initial efforts in the LSR field involved a survey[16] of the shifting and broadening abilities of the entire series of Ln(dpm)$_3$ chelates. In our view a fundamental understanding of the mechanism of action of LSR's can be best achieved through studies of the entire series rather than concentrating on a single member. We chose for our initial study three ligands representative of a variety of classes of organic substrate molecules, namely n-hexanol, 4-picoline N-oxide, and 4-vinylpyridine. The observed shifts were observed to be linear with the parameter Rp in the range 0.03 to 0.252 where Rp is the ratio of the total SR to total substrate concentration. The results of our survey for the most shifted proton resonance of each of these representative ligands are shown in Figure 1. Of the early lanthanides Pr produces the largest upfield shifts while Eu causes downfield displacements of slightly smaller magnitude. Nd produces small upfield shifts while Sm causes extremely small upfield displacements. The latter lanthanides Tb-Yb produce generally larger shifts and group themselves into triads of consecutive elements. Tb, Dy and Ho produce upfield shifts while Er, Tm and Yb cause downfield displacements. In each case the central member (Dy and Tm) produce the largest shifts. Although, owing to the severe signal broadening, spectra are difficult to observe in the presence of Gd(dpm)$_3$, the 4-methyl resonance of 4-picoline was observed in its presence and found to be unshifted.

The absence of an observable shift in the magnetically isotropic $Gd(f^7)$ system is suggestive of a dipolar origin for the shift observed for the other lanthanides. Contact shifts, if present, would be expected to show up in the Gd system. Furthermore, as shown in Table I, there is a perfect correlation between observed isotropic shifts and the signs of magnetic anisotropy data to be found in the literature[17] for several isostructural series of lanthanide complexes. The reasoning here is that if LSR systems represent isostructural series then the magnetic anisotropies will be determined by the particular f^n configuration and, for a given ligand field, will change sign at the same electronic configuration as any other isostructural series. Table I also shows a comparison of typical proton shift data with the quantity g_J-1. For a contact shift mechanism the shifts might be expected to correlate with this quantity (see equation 2). No such correlation is evident and, barring an extremely fortuitous change in sign of (A/h), this provides good evidence that the dipolar mechanism is dominant.

Users of shift reagents, particularly those interested in spectral resolution enhancement, have an interest in the relative degrees of signal broadening produced by the various LSR's. For this reason we carried out an experimental assessment of the broadening abilities of the various dpm chelates. The 2-methyl resonance of 2-picoline was used as a monitor. This signal was chosen because the proximity of these nuclei to the metal amplifies this effect and there is no complication from spin-spin splitting. The results, indicated in Table I, show that although the latter lanthanides are better shifters, they are also more efficient dipolar broadeners. Nevertheless, under conditions of excess substrate, well resolved spectra were obtained for all the systems studied, save those of Gd. For instance, under the conditions of the data in Figure 1, the H-3 resonance of 4-vinylpyridine has an observed width at half height of only 17 Hz for the Dy system. On the other hand it was only for the Pr, Sm, and Eu systems that the 5 Hz spin-spin splitting of the H-2 resonance was observable in the range of Rp values studied. The results suggest that for maximum resolution and significant shifting the Pr, Eu, and Yb systems may be most useful. The extremely potent shifting abilities of Tb, Dy, and Ho may render them of use in specialized circumstances despite their unfavorable broadening characteristics.

III. SOLID STATE STRUCTURAL ASPECTS

The finding that proton isotropic shifts are predominately dipolar in origin is particularly significant in that measurement of these shifts has the potential of yielding detailed geometrical information about the structures and conformations of molecules in fluid solution. Indeed, a large number of investigators have attempted to obtain geometrical information from their studies of LSR systems. Common to most of these studies is the assumption of axial symmetry in which case the second term of eq. 4 vanishes so that the shift ratios correlate with the geometric factor $(3\cos^2\theta-1)r^{-3}$. Correlations have also been attempted with r^{-3} and other distance functions.

Before one accepts this attractive method of structure determination it is important to consider in some detail whether, and under what conditions, the assumption of axial or effective axial symmetry will be valid. Fundamental to our knowledge of solution structure has always been the detailed information regarding solid state structures available from x-ray crystallography. For this reason it is appropriate here to review, very briefly, some of the solid state structures available for LSR systems. Particular attention will be paid to our own work in this area.

In the solid state $Pr(dpm)_3$ has a dimeric structure[18] with 7-coordination achieved for the metal by bridging carbonyl oxygen atoms. On the other hand $Er(dpm)_3$ is monomeric in the crystal[19] with a nearly trigonal prismatic coordination geometry. These findings suggest a low energy of dimerization (it has been estimated at ~10 kcal/mole[20]) and that these chelates can readily expand their coordination spheres. In the presence of substrate ligands adducts are readily formed. Structures of a number of 7-coordinate: $Dy(dpm)_3H_2O$,[21] $Lu(dpm)_3(3\text{-picoline})$,[22] $Eu(dpm)_3(3,3\text{-dimethylthietane 1-oxide})$,[23] $Eu(dpm)_3(dimethyl sulfoxide)$[24] and 8-coordinate: $Ho(dpm)_3(4\text{-picoline})_2$,[25] $Eu(dpm)_3(pyridine)_2$,[26] $Eu(dpm)_3(dimethylformamide)_2$,[27] adducts have been determined. In none of these cases does the substrate molecule occupy a site of rigorous axial symmetry (3-fold or higher rotation axis).

The structure of $Ho(dpm)_3(4-picoline)_2$, which was the first structure of an LSR-organic ligand adduct to be determined,[25] is shown in Figure 2. The coordination sphere is a slightly distorted square antiprism with the picoline ligands occupying corners of opposite square faces as far apart from one another as possible. The complex has crystallographically required C_2 symmetry. It is interesting and significant that when the structure of $Eu(dpm)_3$ (pyridine)$_2$ was solved[26] it was found to have a virtually identical molecular structure, although it crystallizes in a different space group. The picoline ligands do not lie on an axis of symmetry. Indeed, the N-Ho-N angle is 139.1°. Restricting attention to square antiprismatic coordination there are eight geometrical isomers possible for complexes of the $M(O-O)_3S_2$ stoichiometric class, where O-O is a symmetrical bidendate ligand and S is a monodentate substrate. These are shown in Figure 3. We tentatively postulated[25] that the role of bulky substituents in rendering lanthanide chelates efficient as SR's is that internal steric constraints may limit the number of stable geometrical isomers present in solution. If a very large number were present, each with its own magnetic properties and rapidly interconverting on the nmr time scale, dipolar shifts would tend to average toward zero.

The solid state results, taken by themselves, would seem to imply that the assumption of axial symmetry for LSR-substrate complexes in solution is a poor one. The solid state structures, while surely not the only ones present in solution, are likely contributors. On the other hand, an increasing body of evidence is accumulating which demonstrates that good correlations of proton shift ratios with $(3\cos^2\theta-1)r^{-3}$ are obtainable for a variety of rigid ligands. In many cases chemically reasonable values of r and θ are deduced.

Since it is unreasonable to expect all contributing solution structures to be axially symmetric, an alternative explanation for the apparent effective axial symmetry must be sought. One possibility was pointed out by Briggs et al[28] They showed that if a substrate molecule is rotating freely (or with a three-fold or higher symmetry barrier) about an axis which passes through the lanthanide ion, then, even if $\chi_x \neq \chi_y \neq \chi_z$, dipolar shifts will be given by eq. 6.

$$\frac{\Delta H^{dip}}{H} = \frac{1}{2}[-D(3\cos^2\alpha-1)-D'(\sin^2\alpha\cos2\beta)](3\cos^2\gamma-1)r^{-3} \quad (6)$$

where γ is the angle made by the radius vector to the resonating nucleus with the axis of rotation, and α and β are the spherical polar angular coordinates of the rotational axis analogous to θ and Ω of eq. 4. A detailed analysis of our solid state structure[25] suggests that severe ortho-hydrogen – carbonyl oxygen steric interactions render free rotation of the picoline ligands impossible if the remainder of the structure is left unchanged.

An alternative possibility,[24] which might lead to effective axial symmetry, is that there are a reasonably large number of contributing structures in solution and the non-axial portion of the dipolar field, as seen by a given substrate, is effectively eliminated in time average. It should be remembered that the extremely short electronic spin-lattice relaxation times exhibited by most of the lanthanide ions enable the magnetic properties to follow instantaneously any nuclear displacements. Further work is clearly necessary to resolve this important point.

IV. DIRECT EVALUATION OF DIPOLAR SHIFTS FROM MAGNETIC ANISOTROPY DATA

While the results of many investigations such as those discussed in Section II indicate a predominately dipolar origin for the observed proton resonance shifts in LSR systems, we deemed it worthwhile to obtain direct, independent evidence of this by means of single crystal magnetic anisotropy studies. Complexes of the type $Ln(dpm)_3$ (4-picoline)$_2$ readily form large, reasonably air stable, crystals which are isomorphous for the entire lanthanide series. Since the structure of the Ho complex has been determined,[25] this series was chosen for our investigation.[29] The principal crystal anisotropies were measured by Krishnan's critical torque method, the theory, methodology and apparatus of this technique is described in detail elsewhere.[30] These data, when combined with the average susceptibilities, $\bar{\chi}$, measured by Gouy or nmr methods, yield the principal crystal susceptibilities listed in Table II. The direction of the y molecular magnetic axis is coincident with the C_2 (b) crystal axis, but the directions of the x and z molecular axes are not specified by symmetry and cannot be determined from the anisotropy experiment. The required relationship may be expressed by a parameter η, the angle the molecular z axis makes with the crystal $\underline{a}$ axis. Table II lists the molecular anisotropy factors D and D' derived for $\eta = 18°$, a value in the range of η yielding the best fits for the dipolar shifts of the 4-methyl resonance of 4-picoline for the entire series of adducts. The marked magnetic nonaxiality of the various adducts is evident. Dipolar shifts for the freely rotating methyl group evaluated from the magnetic and structural[25] data using eq. 4 for η in the range 2° to 22° are shown in Figure 4 along with the experimental nmr results. The assumption is made here that the solid state structure persists in solution, retains its solid-state magnetic properties, and is the only species present. In reality this is not expected to be entirely true, but the degree of agreement suggests that this is a reasonable first approximation and demonstrates the basic dipolar nature of the shifts. With the exception of the Tm system for which the agreement is poorest, the second term of eq. 4 contributes less than 15 percent to any of the shifts. This is not expected to be a general result, but occurs because the axial geometric factor is greater by a factor of about 26 than the equatorial one in the present case. It should be noted that while the exact value of the parameter η is unknown, the quality of agreement is not particularly sensitive to its value. Indeed, qualitative agreement (direction of shifts) is achieved for all the complexes (save those of Tm) for any η in the range 0 to 30°.

Taken in conjunction with the structural data outlined in Section III, our results show that magnetic axiality in lanthanide complexes cannot be assumed in general and suggest that detailed structural inferences based on this assumption must be accepted with reservation. These considerations will be particularly important in metal-loproteins where the metal ions occupy sites of low symmetry and ligand (polypeptide) contraints prevent the most likely mechanisms for achieving effective axial symmetry (see discussion in the previous section).

V. THEORY

To be useful, any theory of dipolar shifts in LSR's must be capable of accounting for the variation in signs and magnitudes of the magnetic anisotropies and hence of the observed shifts as one progresses across the series from Pr to Yb. Ideally, such a theory should be capable of applying and testing a variety of ligand field models and be able to handle the low-symmetry situations usually found in shift reagent systems.

Quite recently, Bleaney[15] has presented a theoretical treatment of the variation of magnetic anisotropies across the series. Restricting his attention to systems of axial symmetry he derived concise expressions for the magnetic anisotropies to be expected. These expressions correlate in a semiquantitative manner with proton resonance shifts observed for several series of lanthanide complexes and LSR systems.[31]

While it has the advantage of being concise, this theory has several flaws, mostly in regard to unmet beginning assumptions. First of all, it assumes that the first order Zeeman (FOZ) contribution to the anisotropy vanishes at the temperatures of the usual nmr experiments. As we will demonstrate below, this is not true in general, although it will occur for low-symmetry, even-electron systems. Secondly, Bleaney's theory predicts a linear T^{-2} temperature dependence of the observed shifts. Neither our experimental nor our theoretical results are in accord with this prediction (vide infra). Finally, Bleaney treats the magnetic anisotropies as proportional to the phenomenological crystal field parameters $A_2^0 \langle r^2 \rangle$ and $A_2^2 \langle r^2 \rangle$. Our more general treatment shows that while the $A_2^0 \langle r^2 \rangle$ term is important when incorporated into the calculation of magnetic anisotropies, it and $A_2^2 \langle r^2 \rangle$ cannot be labeled as the sole contributors to the variation in magnetic anisotropies.

Our theoretical approach, which is a general one readily extendable to systems of any symmetry is a direct outgrowth of our treatment of the magnetic properties of transition metal complexes.[32,33] It has the further advantage that it can apply and test a variety of ligand field models. Our calculational procedure encompasses the entire ground term of each f^n configuration rather than

concentrating on individual J levels. All of the J mani-
folds of a particular term are included in the calculation.
This inclusion of higher J states is particularly important for complexes of Eu^{3+} (f^6) where the ^7F term includes
J=0,1,2,3,4,5,6, with the diamagnetic J=0 level lying
lowest. The paramagnetic J=1 and J=2 levels are appreciably thermally populated at room temperature and must be
included in any valid treatment.

In performing the energy and magnetic calculations we
employ a $|L,S,M_L,M_S\rangle$ basis rather than the more usual
$|L,S,J,M_J\rangle$ basis. For illustration, we will outline our
procedure for a Pr(III) complex (f^2, ground term:
$^3H_{J=4,5,6}$). First the "ligand field" (LF) matrix elements
of the type $\langle m_\ell | V_{LF} | m'_\ell \rangle$ one-electron f orbitals are
calculated by one of the following three methods.

(1) <u>Phenomenological Crystal Field Theory</u> - (a) Matrix elements may be calculated in terms of the purely
phenomenological parameters $A_\lambda^\kappa \langle r^\lambda \rangle$ <u>via</u> the operator equivalent techniques[34,35] or through the use of irreducible
tensor operators[36-39]
(b) Matrix elements may be computed from an electro-
static point charge model.[32,33] In this case the para-
meters are q_i, the charge on the ith ligand, Θ_i and Φ_i,
the angular spherical polar coordinates of the ith ligand
and ρ_2, ρ_4, ρ_6, the second, fourth and sixth order radial
parameters ($\rho_\lambda = r^\lambda / R^{\lambda+1}$ where R represents the average metal
to point charge distance.) While the ρ_λ and q_i parameters
are sometimes ascribed physical significance, they too are
more realistically viewed as empirical parameters.

(2) <u>Orbital Overlap Model.</u> - This is our version of
the angular overlap model introduced by Jørgensen <u>et al.</u>[40]
and amplified by Schäffer.[41] In this theory f orbital
splitting is attributed to weak covalent bonding which is
proportional to metal-ligand overlaps, resulting in different antibonding energies for the various f orbitals.
We employ standard programs for machine computation of
overlap integrals[42] (both radial and angular parts) rather
than hand calculation of the angular portions alone as
was performed previously.[40,41]

(3) Effective Perturbation Method. - This method, recently proposed by one of us,[43] allows one to take the results of one-electron MO calculations to obtain LF matrix elements for use in many-electron LF calculations. The parameters of the method are the MO energies and in some cases the MO eigenfunctions of the "mainly f" orbitals.

With the 7x7 matrix of one-electron LF matrix elements as raw material our calculation proceeds as follows. The $|L,S,M_L,M_S\rangle$ basis functions are expressed as Slater determinantal wave functions and matrix elements between these are readily given in terms of the one-electron matrix elements. For the LF part of the ^3H ground term of Pr(III) this results in an 11x11 matrix. In our $|L,S,M_L,M_S\rangle$ basis the spin-orbit coupling matrix (33x33) is readily written down. The 11x11 LF and the 33x33 spin-orbit matrices are added appropriately (LF matrix elements connect only states with the same M_S values.) Diagonalization of the resulting 33x33 matrix yields the desired energies and eigenfunctions. Magnetic properties are then calculated using Van Vleck's equation. All of these operations are achieved through use of a computer program developed by us. For the Pr(III) example with a spin-orbit coupling constant $\lambda = 380$ cm^{-1} the calculation yields for a typical example (ethylsulfate) a ground J=4 manifold of 9 levels in the 0-250 cm^{-1} range, a J=5 manifold (11 levels) in the 1900-2100 cm^{-1} range and a J=6 manifold (13 levels) in the 4100-4500 cm^{-1} range.

While our method is conceptually simple, it does involve diagonalization of large matrices, which would be impossible without high speed, large-core digital computation. Working versions of our programs have been developed for the entire lanthanide series with the exception of the orbitally non-degenerate f^7 (^8S) configuration.

In order to achieve a direct comparison with the treatment by Bleaney in the prediction of magnetic anisotropy trends we have initially concentrated on employing the results of purely phenomenological crystal field theory. Chachra and Mookherji[44] have shown that a model with parameters $A_2^0 \langle r^2 \rangle = 100$, $A_4^0 \langle r^4 \rangle = -55$, $A_6^0 \langle r^6 \rangle = -35$, and $A_6^{\pm6} \langle r^6 \rangle = 510$ cm^{-1} reproduces the magnetic anisotropy of ytterbium ethylsulfate over the temperature range 100-300°K. Lanthanide ethylsulfates represent a well-studied, isomorphous class of axially symmetric 9-coordinate aquo-complexes. The 9-coordination is achieved by a trigonal prism of water oxygen atoms with three additional water oxygens arranged in the equatorial plane and rotated through $\pi/3$ with respect to the prism triangles.[45] The equatorial oxygens are slightly farther away from the central metal ion than the prismatic ones; the complexes thus possess C_{3h} point symmetry.

Using the Chachra and Mookherji parameters, the energy level diagram shown in Figure 5 was calculated. The ground $J=7/2$ manifold consists of four Kramers doublets lying in the range 0 to 240 cm^{-1}. It should be noted that their energies are not all $\ll kT$ at room temperature (~ 200 cm^{-1}) and that thermal population of excited states is by no means negligible. The g-values for the various levels are also indicated in Figure 5, and the values calculated for the lowest Kramers doublet are in good agreement with those determined experimentally by liquid helium magnetic susceptibility measurements.[46] This model also accurately predicts a first excited level at 40.8 cm^{-1} which has been observed at 42.0 cm^{-1} spectroscopically.[47]

The calculated $\Delta\chi = \chi_{\parallel} - \chi_{\perp}$ values plotted vs. T^{-1} are shown in Figure 6 along with the experimental points. The quality of agreement of this phenomenological model with a variety of spectroscopic and magnetic data suggests that it is a reasonably accurate one. Several features emerge from an analysis of our magnetic calculations. At higher temperatures the calculated $\Delta\chi$ values follow a very good T^{-1} relationship (such T^{-1} plots would have to be concave upwards in order for T^{-2} plots to produce linear graphs). The second order Zeeman (SOZ) contribution to $\Delta\chi$, $\Delta\chi(SOZ)$, is dominant and linear in T^{-1}, but is counteracted by a non negligible $\Delta\chi(FOZ)$ which is opposite in sign. None of these findings are in agreement with Bleaney's[15] theory.

As an initial step in accounting for the variation of the magnetic anisotropies across the lanthanide series we have chosen to keep the ligand field fixed and have used the above-mentioned ytterbium ethylsulfate parameters to calculate the elements of the 7x7 LF matrix over the complex f orbital basis. This same 7x7 LF matrix is used to calculate the magnetic properties of all the f^n configurations. The only remaining parameters are the spin-orbit coupling constant values which have either been taken from the literature[48] or deduced from the spectra of the particular ethylsulfate. This procedure is analogous to Bleaney's assumption of a constant $A_2^0 \langle r^2 \rangle$ value across the series.

The calculated energy level diagrams including g-values and thermal populations at 300°K are shown for the f^2, and f^6 configurations in Figures 7 and 9, respectively. The corresponding $\Delta\chi$ values plotted vs. T^{-1} are shown in Figures 8 and 10 respectively, where the contribution of FOZ and SOZ effects are indicated. From inspection of these graphs and a summary of some of these quantities for some other configurations as well,   given in Table III, a number of generalizations can be made. At room temperature there is always a number of thermally populated excited states with energies ranging from zero to ∿2kT in some cases, with irregular spacings. For even electron systems, since $g_\perp = 0$ for all levels, $\Delta\chi$(FOZ) is always positive. For the ethylsulfate system this quantity is likewise positive for the odd electron systems. The FOZ contributions to the anisotropies are definitely not generally negligible. Most importantly, the signs of the total anisotropies correlate perfectly with the signs of the observed isotropic shifts for the isostructural series. Their magnitudes are in semi-quantitative agreement with those of the observed shifts. Space does not permit a detailed analysis of all of the configurations, but a few of the more important features are worthy of mention.

Pr(III), f^2. (Figures 7 and 8). The plot of $\Delta\chi$ vs. T^{-1} is reasonably linear over a fairly wide range with perhaps a small T^{-2} component at the highest temperatures. Below 100°K $\Delta\chi$ goes through a maximum. It is unlikely that this sort of complex behavior can be observed in dipolar shifts since the temperatures available to solution nmr studies normally restrict T^{-1} to 3-4 x 10^{-3}°K^{-1} range.

Eu(III), f^6. (Figures 9 and 10). This is a particularly interesting case in that the ground state is non-magnetic (J=0), but interacts with low-lying thermally populated excited states (J=1,2). At higher temperatures $\Delta\chi$ is quite linear in T^{-1}, but, as found in many other lanthanide systems, the plot has a large non-zero intercept on the T^{-1} axis. At low temperatures the anisotropy becomes constant and is entirely due to the SOZ effect. Again the interesting (non T^{-1}) behavior occurs at temperatures below those available to solution nmr studies. The temperature dependence of $\Delta\chi$ predicted by our calculation is exactly reproduced by the analytical function given by Bleaney.[15]

In order to determine the sensitivity of the calculated anisotropies to the various phenomenological crystal field parameters, calculations were performed in which $A_2^0 \langle r^2 \rangle$, $A_4^0 \langle r^4 \rangle$, $A_6^0 \langle r^6 \rangle$, and $A_6^6 \langle r^6 \rangle$ were each increased by 50% in absolute magnitude. The results set out in Table IV show that the $A_2^0 \langle r^2 \rangle$ term is by far the most sensitive parameter, but others, in particular $A_4^0 \langle r^4 \rangle$, have a non-negligible effect.

Table IV

Percentage Change in $\Delta\chi$ Upon Increasing Crystal Field Parameter $A_\lambda^K \langle r^\lambda \rangle$ by 50%

Ion	f^n	$\Delta\chi$(total)	A_2^0	A_4^0	A_6^0	$A_6^{\pm6}$
Ce^{3+}	f^1	793	+49	+27	−3	+6
Pr^{3+}	f^2	876	+62	−11	−5	−12
Eu^{3+}	f^6	−583	+51	−4	+4	−5
Tb^{3+}	f^8	15742	+52	+6	−5	+5
Tm^{3+}	f^{12}	−6983	+37	+7	0	+6
Yb^{3+}	f^{13}	−1829	+42	−10	+4	−4

It is clear to us that, while the phenomenological crystal field theory may provide a reasonable description of magnetic and electronic properties in systems of relatively high symmetry where a minimum number of parameters (four in the case of the axial ethylsulfates) may adequately describe the ligand field, extension to systems of lower symmetry is not feasible owing to the larger number of physically meaningless parameters in the expansion. For this reason we are exploring alternative models for calculating ligand field matrix elements over the one-electron f orbital basis.

While a generally satisfactory LF model is not available presently, our approach is shown to be capable of yielding an understanding of the anisotropy trends observed across the lanthanide series. The calculations are of such a nature that, providing an adequate LF model is at hand, a detailed analysis of the magnetic properties of the ground and excited states of lanthanide complexes is possible.

VI. EXPERIMENTAL TEMPERATURE DEPENDENCE OF MAGNETIC ANISOTROPY

We were prompted by Bleaney's[15] prediction of a T^{-2} temperature dependence, as well as by our own assessment of the theory, to measure the magnetic anisotropies of some representative shift reagent system in order to obtain more insight into the actual behavior of these complexes. We thus chose several of the $Ln(dpm)_3(4-picoline)_2$ systems whose single crystal magnetic anisotropy studies are discussed in Section IV, namely Ln = Pr, Nd, Eu, and Yb. As previously mentioned, these systems are not axially symmetric, nor are the directions of two of the magnetic axes known. We will deal, therefore, directly with the crystal rather than the molecular anisotropies since they represent experimental observables. As indicated in Section V, since a theoretical model capable of dealing with such low symmetry systems has not yet been developed, we must be content here with an exposition of some experimental facts as detailed theoretical interpretation awaits further work.

Our single crystal anisotropy data plotted vs. T^{-1} in the 70-300°K range are shown for Pr(III), and Eu(III) in Figures 11 and 12, respectively. (In these figures the subscript of $\Delta\chi$ designates the axis of suspension in the critical torque experiment.)

The temperature dependencies are in general not simple. Certain ones exhibit excellent T^{-1} dependencies while others show varying degrees of curvature. A few pass through maxima with $\Delta\chi_a$ of the Pr(III) system

67

vanishing before changing sign just above liquid nitrogen
temperatures. These maxima are reminiscent of the maximum
calculated for the axially symmetric praseodymium ethylsulfate system. Both the anisotropies and the principal
crystal susceptibilities of the Eu system appear to be
approaching constant values at lower temperatures as would
be expected for a J=0 ground state with a large SOZ contribution to the susceptibility. It should be emphasized
that these results suggest that reasonably good T^{-1} plots
may be expected over the limited temperature range available to dipolar shift experiments, but that such plots
will not in general pass through the origin.

VI. CONCLUSIONS

The predominately dipolar nature of proton resonance
shifts in LSR systems has been established. The signs
and magnitudes of the shifts are predictable from the
results of single crystal magnetic anisotropy experiments.
The available evidence suggests that the assumption of
structural and/or magnetic axiality for LSR systems in
solution is not a good one, although there are certain
mechanisms by which effective axial symmetry may be achieved. The problems of magnetic nonaxiality and nonneglig-
ible contact contributions, for which there is now compelling evidence particularly for nuclei other than hydrogen,
suggest that detailed structural inferences drawn from
LSR data be accepted with some caution.

A published theory[15] of the variation of magnetic
anisotropy across the lanthanide series has been critically
examined and found to have several flaws. We offer an
alternative, more general theoretical approach which accounts for the observed trends. Furthermore, it may be
extended to low-symmetry (nonaxial) systems.

The temperature dependencies of the $\Delta\chi$'s have been
measured for several LSR adducts and found not to be simple.
The temperature dependencies are definitely not linear in
T^{-2}. Our theoretical results are consistent with these
findings.

Acknowledgement. This research was supported by the
National Science Foundation through grant GP 26148.

REFERENCES

1. "NMR in Paramagnetic Molecules: Principles and Applications", G. N. La Mar, W. DeW. Horrocks, Jr. and R. H. Holm eds, Academic Press Inc., New York, 1973 in press.

2. C. C. Hinckley, J. Amer. Chem. Soc., 91, 5160 (1969).

3. W. DeW. Horrocks, Jr., Chapter 12 of ref. 1.

4. J. Reuben in "Progress in NMR Spectroscopy", J. W. Emsley, J. Feeney, and L. H. Sutcliffe eds, Pergamon Press, Oxford 1973 in press.

5. W. DeW. Horrocks, Jr., R. C. Taylor, and G. N. La Mar, J. Amer. Chem. Soc., 86, 3031 (1964).

6. G. N. La Mar, W. DeW. Horrocks, Jr. and L. C. Allen, J. Chem. Phys., 41, 2126 (1964).

7. R. W. Kluiber and W. DeW. Horrocks, Jr., J. Amer. Chem. Soc., 87, 5350 (1965).

8. W. DeW. Horrocks, Jr., R. H. Fischer, J. R. Hutchison, and G. N. La Mar, J. Amer. Chem. Soc., 88, 2436 (1966).

9. G. N. La Mar, R. H. Fischer, and W. DeW. Horrocks, Jr. Inorg. Chem., 6, 1798 (1967).

10. W. DeW. Horrocks, Jr., Inorg. Chem., 9, 690 (1970).

11. W. DeW. Horrocks, Jr. and E. S. Greenberg, Inorg. Chem. 10, 2190 (1971).

12. W. DeW. Horrocks, Jr. and D. DeW. Hall, Inorg. Chem., 10, 2368 (1971).

13. R. J. Kurland and B. R. McGarvey, J. Magn. Reson., 2, 286 (1970).

14. B. R. McGarvey, J. Chem. Phys., 53, 86 (1970).

15. B. Bleaney, J. Magn. Reson., 8, 91 (1972).

16. W. DeW. Horrocks, Jr. and J. P. Sipe, III, J. Amer. Chem. Soc., 93, 6800 (1971).

17. See ref. 16 for citations.

18. C. S. Erasmus and J. C. A. Boeyens, Acta Crystallogr. Sect. B, 26, 1843 (1970).

19. J. P. R. de Villiers and J. C. A. Boeyens, Acta Crystallogr.Sect. B, 27, 2335 (1971).

20. J. C. A. Boeyens, J. Chem. Phys., 54, 75 (1971).

21. C. S. Erasmus and J. C. A. Boeyens, J. Cryst. Mol. Struct., 1, 83 (1971).

22. S. J. S. Wasson, D. E. Sands, and W. F. Wagner, Inorg. Chem., 12, 187 (1973).

23. J. J. Uebel and R. M. Wing, J. Amer. Chem. Soc., 94, 8910 (1972).

24. See D. S. Dyer et al., this volume.
25. W. DeW. Horrocks, Jr., J. P. Sipe, III, and J. R. Luber, J. American Chem. Soc., 93, 5258 (1971).
26. R. E. Cramer and K. Seff, J. Chem. Soc., Chem. Commun., 400 (1972).
27. See ref. 24.
28. J. M. Briggs, G. P. Moss, E. W. Randall, and K. D. Sales, J. Chem. Soc. Chem. Commun., 1180 (1972).
29. W. DeW. Horrocks, Jr. and J. P. Sipe, III, Science, 177, 994 (1972).
30. W. DeW. Horrocks, Jr. and D. DeW. Hall, Coord. Chem. Rev., 6, 147 (1971).
31. B. Bleaney, G. M. Dobson, B. A. Levine, R. B. Martin, R. J. P. Williams, and A. V. Xavier, J. Chem. Soc. Chem. Commun., 791 (1972).
32. W. DeW. Horrocks, Jr. and E. S. Greenberg, Abstracts 162nd ACS National Meeting, Washington, D.C. September 1971 paper INOR 189.
33. E. S. Greenberg, Ph.D. Thesis, The Pennsylvania State University, 1973.
34. R. J. Elliott and K. W. H. Stevens, Proc. Roy. Soc. (London), A219, 387 (1953).
35. B. R. Judd, Proc. Roy. Soc. (London), A227, 552 (1955).
36. B. G. Wybourne, "Spectroscopic Properties of Rare Earths," Interscience, New York, 1965.
37. V. Fano and G. Racah, "Irreducible Tensorial Sets," Academic Press, New York, 1959.
38. B. R. Judd, "Operator Techniques in Atomic Spectroscopy," McGraw-Hill, New York, 1963.
39. J. S. Griffith, "The Irreducible Tensor Method for Molecular Symmetry Groups," Prentice-Hall, Englewood Cliffs, New Jersey, 1962.
40. C. K. Jørgensen, R. Pappalardo, and H. H. Schmidtke, J. Chem. Phys., 39, 1422, (1963).
41. C. E. Schäffer, Struct. Bonding (Berlin), 5, 68 (1968).
42. R. Hoffmann, J. Chem. Phys., 39, 1397 (1963).
43. W. DeW. Horrocks, Jr., J. Amer. Chem. Soc., 94, 656 (1972).
44. A. Mookherji and S. P. Chachra, Ind. J. Pure Appl. Phys., 7, 560 (1969).
45. D. R. Fitzwater and R. E. Rundle, Zeit. Krist., 112, 362 (1959).
46. A. H. Cooke, F. R. McKim, H. Mayer and W. P. Wolf, Phil. Mag., 2, 928 (1957).

47. E. Y. Wong, *J. Chem. Phys.*, 39, 2781 (1963).
48. G. H. Dieke, "Spectra and Energy Levels of Rare Earth Ions in Crystals," Interscience, New York, 1968.

Table I

Isotropic Shift and Linewidth Data for Ln(dpm)$_3$ - Picoline Systems, Correlation with Magnetic Anisotropy and g_J-1.

Ln	$\Delta H^{iso}/H^a$	$\Delta\upsilon_{1/2}$, Hz^b	$\chi_\parallel - \chi_\perp{}^c$ Ln(C$_2$H$_5$SO$_4$)$_3\cdot$9H$_2$O	LnNa(MoO$_4$)$_2$	g_J-1
Pr	6.2	5.6	+		-1/5
Nd	3.0	4.0	+		-3/11
Sm	0.7	4.4	+		-5/7
Eu	-3.0	5.0			(1/2)^d
Tb	31.7	96		+	1/2
Dy	43.2	200	+	+	1/3
Ho	23.8	50	+	+	1/4
Er	-10.4	50	-	-	1/5
Tm	-22.0	65		-	1/6
Yb	-10.9	12		-	1/7

^{a}In ppm for H-2 of 4-picoline measured at 30° with Rp = 0.125
^{b}Width at half maximum for methyl resonance of 2-picoline.
^{c}From ref. 17.
^{d}For first excited J=1 state.

Table II

Principal Crystal Susceptibilities and Molecular Anisotropy Factors for Ln(dpm)$_3$(4-picoline)$_2$ Systems at 298°K.[a]

Ln	χ_a	χ_b	χ_c	$\chi_z - \frac{1}{2}(\chi_x + \chi_y)$	$\chi_x - \chi_y$
Pr	2779	6315	4865	−3180	−1204
Nd	3849	5903	4337	−1357	−1508
Sm	931	1249	973	−188	−271
Eu	5606	3610	4861	1502	1163
Tb	28268	62209	26589	−15834	−35818
Dy	33033	65868	45114	−24597	−19328
Ho	38811	54796	41768	−9995	−12680
Er	35984	27572	36801	3653	9326
Tm	22688	9072	30051	1823	21848
Yb	9102	3563	8093	3453	4401

[a] In VVk/mole, see ref. 30.

Table III

Calculated Susceptibility Anisotropies at 300°K Using
Ytterbium Ethylsulfate Ligand Field Parameters.[a,b]

Ion	f^n	Term	J	$\Delta\chi$(FOZ)	$\Delta\chi$(SOZ)	$\Delta\chi$(Total)	Sign of $\Delta H^{iso}/H^c$
Ce^{3+}	f^1	^2F	5/2	2515	-1721	793	
Pr^{3+}	f^2	^3H	4	715	161	876	+
Nd^{3+}	f^3	^4I	9/2	772	-696	76	+
Pm^{3+}	f^4	^5I	4	436	-671	235	
Sm^{3+}	f^5	^6H	5/2	270	-213	57	+
Eu^{3+}	f^6	^7F	0,1,2	1467	-2050	-583	−
Tb^{3+}	f^8	^7F	6	49853	-34111	15742	+
Tm^{3+}	f^{12}	^3H	6	3937	-10920	-6983	−
Yb^{3+}	f^{13}	^2F	7/2	1978	-3807	-1829	−

[a]Owing to computational problems at press time calculations
were not available for the f^9, f^{10}, and f^{11} configurations.

[b]$\Delta\chi = \chi_\| - \chi_\perp$.

[c]From ref. 16.

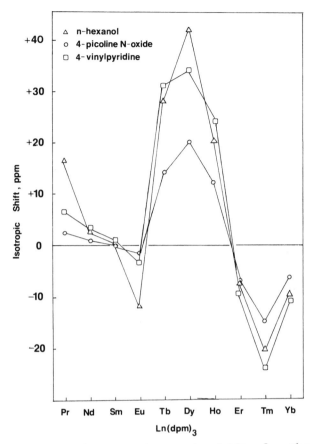

Figure 1. Observed proton isotropic shifts for the most shifted resonances of deuteriochloroform solutions of n-hexanol (H-1), 4-picoline N-oxide (H-2), and 4-vinyl-pyridine (H-2) in the presence of $Ln(dpm)_3$ with R_p = 0.125 at 30°C. From reference 16.

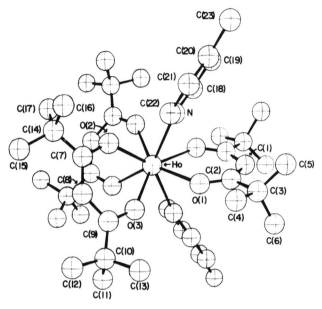

Figure 2.   Molecular structure of Ho(dpm)$_3$(4-picoline)$_2$.

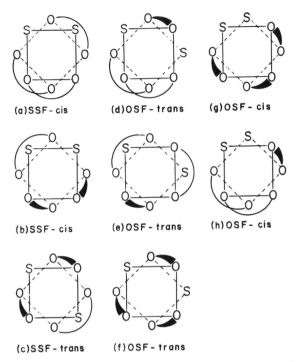

(a)SSF-cis (d)OSF-trans (g)OSF-cis

(b)SSF-cis (e)OSF-trans (h)OSF-cis

(c)SSF-trans (f)OSF-trans

Figure 3. Geometrical isomers possible for complexes of the stoichiometry $M(O-O)_3S_2$ based on square antiprismatic coordination.

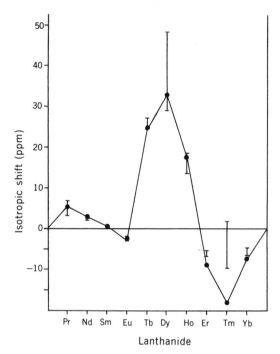

Figure 4. Comparison of the observed isotropic shifts (298°K) for the 4-picoline methyl resonance (points connected by solid line) in the $Ln(dpm)_3(4-picoline)_2$ systems with the dipolar shifts evaluated from single-crystal magnetic anisotropy data for values of the parameter η in the range 2° to 20° (see text), from reference 29.

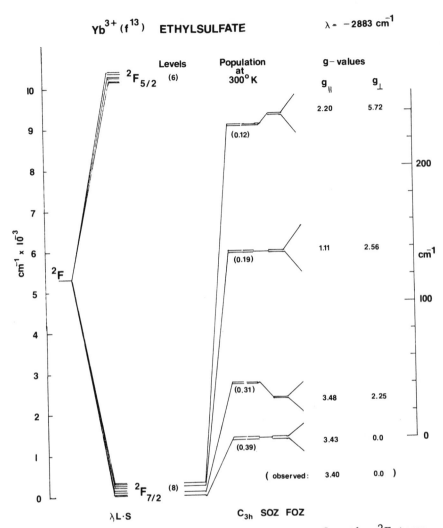

Figure 5. Calculated energy level diagram for the ^2F term of Yb^{3+}, f^{13}, using Yb^{3+} ethylsulfate parameters.

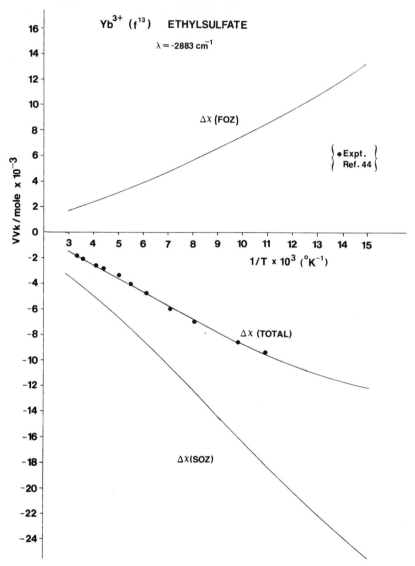

Figure 6. Calculated temperature dependence of $\Delta\chi$ for the Yb^{3+} system.

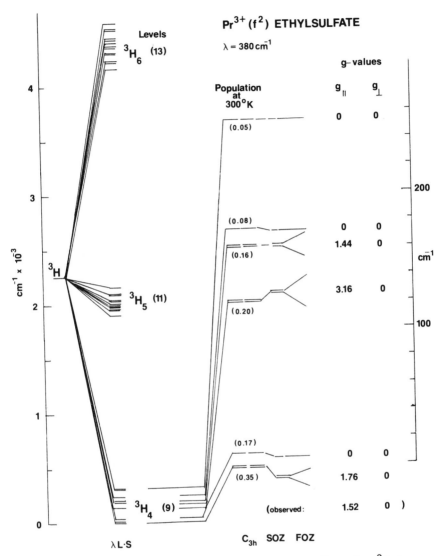

Figure 7. Calculated energy level diagram for the ^{3}H term of Pr^{3+}, f^2, using Yb^{3+} ethylsulfate parameters.

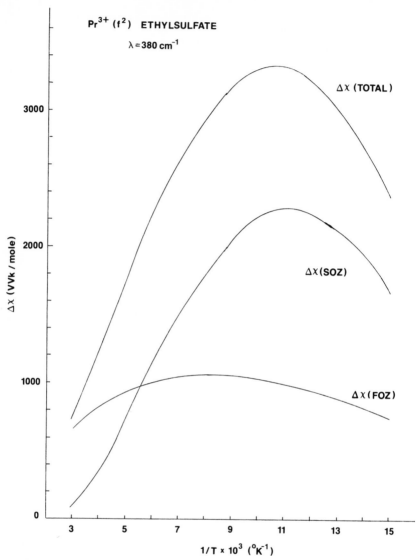

Figure 8. Calculated temperature dependence of $\Delta\chi$ for the Pr^{3+} system.

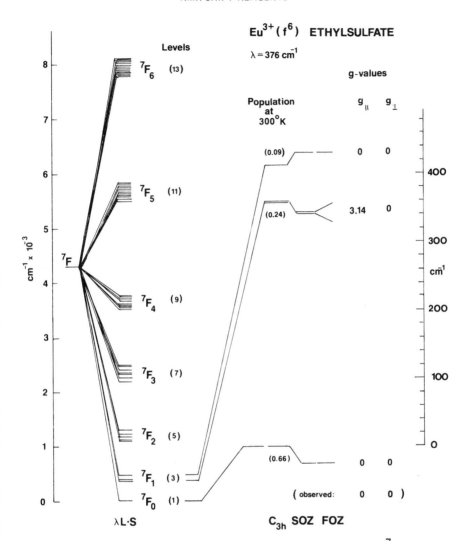

Figure 9. Calculated energy level diagram for the ^7F term of Eu^{3+}, f^6, using Yb^{3+} ethylsulfate parameters.

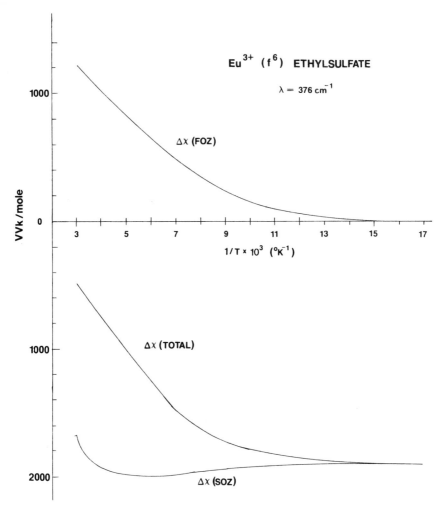

Figure 10. Calculated temperature dependence of $\Delta\chi$ for the Eu^{3+} system.

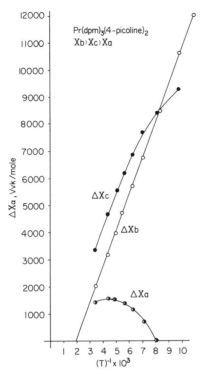

Figure 11. Temperature dependencies of the principal crystal anisotropies of Pr(dpm)$_3$(4-picoline)$_2$. The subscript to $\Delta\chi$ indicates the axis of suspension. The $\underline{b}$ axis is parallel to the C$_2$ axis of the complex.

85

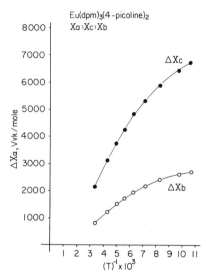

Figure 12. Temperature dependencies of the principal crystal anisotropies of $Eu(dpm)_3(4\text{-picoline})_2$. See caption to Figure 11.

CHIRAL SHIFT REAGENTS[*]

Charles Kutal
Department of Chemistry
University of Georgia
Athens, Georgia

INTRODUCTION

The introduction of chiral lanthanide shift reagents has greatly facilitated the determination of enantiomeric purity by NMR. The method is based upon the premise, first verified by Pirkle for optically active solutes in chiral solvents (1), that the NMR spectra of enantiomers are nonequivalent in a dissymmetric environment. Since the magnitude of the lanthanide induced shift, $\Delta\delta$, has proven to be a sensitive probe of steric environment (2), it is expected that chiral shift reagents C_+ (where + denotes one optical isomer of the chelate) will produce a differential shift, $\Delta\Delta\delta$, between the resonances of equivalent nuclei in an enantiomeric pair (R, S). The situation is described by eqn. 1, where the

$$C_+ \; + \; R \; \rightleftarrows \; (C_+ R)$$
$$C_+ \; + \; S \; \rightleftarrows \; (C_+ S) \tag{1}$$

association of C_+ with R and S forms two diastereomeric (and, in principle, NMR distinguishable) adducts in solution. Because the interchange processes are rapid,

[*]Editor's Note: At the Dallas A.C.S. Symposium the paper on chiral shift reagents was presented by Professor H. L. Goering. Due to a death in his family, Professor Goering was not able to prepare a chapter. Since the topic is one of both interest and importance, Dr. Kutal was invited to write a review of the work in this field.

the observed shifts are a time average of those for the complexed and free substrate. Thus $\Delta\Delta\delta$ and $\Delta\delta$ are dependent on the molar shift reagent/substrate ratio. The intensity of a signal, on the other hand, is proportional to the total mole fraction, n, of a given enantiomer. From the measurement of peak areas, the enantiomeric purity can be readily calculated using eqn. 2.

$$E.\,P. \;=\; \frac{n_R - n_S}{n_R + n_S} \qquad (n_R \geq n_S) \qquad (2)$$

The analogous technique in which chiral solvents are employed remains a viable alternative for determining enantiomeric purity in certain cases. Chiral shift reagents, however, are generally found to be effective with a wide variety of organic Lewis bases. In addition, they normally produce much larger chemical shift differences between enantiomers and are thus valuable in the interpretation of complex spectra.

Table 1 lists several lanthanide shift reagents (each identified by a Roman numeral) along with references to their preparation. The chiral complexes are discussed in the first part of the next section; the second part briefly deals with a related topic, namely, the utility of achiral shift reagents in distinguishing diastereotopic nuclei (3).

EXPERIMENTAL STUDIES

Chiral Shift Reagents

The feasibility of using chiral shift reagents was first demonstrated by Whitesides and Lewis (4), who observed nonequivalent spectra for enantiomeric amines in the presence of tris[3-(tert-butylhydroxymethylene)-d-camphorato]europium(III), I. Figure 1 illustrates the effect of I on the spectra of 1-phenylethylamine. The proton resonances of the (S) enantiomer exhibit the expected downfield pseudocontact shift due to the interaction with the paramagnetic europium(III) ion. In a mixture of enantiomers, however, chemical shift differences ranging from ~0.5 ppm for the C$\underline{H}$NH$_2$ proton to ~0.07 ppm for the para hydrogens of the benzene ring are clearly evident. Differential shifts were also observed for the C$\underline{H}$NH$_2$ resonances of (R)- and (S)- amphetamine ($\Delta\Delta\delta$ = 0.7 ppm) and the C$\underline{H}_2$CHNH$_2$ resonances of (R)- and (S)-aminobutane ($\Delta\Delta\delta$ = 1.4 ppm). Less basic substrates such as 2-octanol and 1-phenylethanol exhibited values of $\Delta\Delta\delta \cong 0$, even though their spectra displayed appreciable pseudocontact shifts.

In a later study (5), a series of new chiral shift reagents was reported (II-VII). Considered together, they were effective for several substrates including amines, alcohols, ketones, esters, and sulfoxides. No single reagent was superior to the others for each class of compound, however.

The observation that fluorinated lanthanide chelates are stronger Lewis acids and more effective shift reagents (6,7) prompted the use of tris[3-(trifluoromethyl-hydroxymethylene)-$\underline{d}$-camphorato]europium(III), VIII (also called tris(trifluoroacetyl-$\underline{d}$-camphorato)europium(III), abbreviated facam), tris[3-(heptafluoropropylhydroxy-methylene)-$\underline{d}$-camphorato]europium(III), IX, and their praseodymium(III) analogs X and XI (8-10). As seen from the compilation of data in Table 2, these complexes are generally superior to nonfluorinated shift reagents in producing larger $\Delta\Delta\delta$ values. Fraser, et. al. (10) found that use of a fluorinated shift reagent permitted differentiation in all seven pairs of enantiomers studied while the non-fluorinated reagent was ineffective for five of the seven pairs. Figures 2a and 2b contain the spectra of d,l-2-phenyl-2-butanol in the presence of achiral tris-(dipivalolymethanato)europium(III), XII and the chiral VIII, respectively. While both reagents induce comparable pseudocontact shifts, VIII effects a separation of the corresponding α-methyl singlets by 0.29 ppm and th β-methyl triplets by 0.22 ppm or ~2J, resulting in a quintuplet. In Figure 2c the spectrum of partially resolved 1,2-dimethyl-exo-2-norbornanol in the presence of VIII exhibits a separation of the analogous methyl groups in each enantiomer. Dongala, et. al. (11) have reported using VIII to determine the enantiomeric purity of several β-hydroxyesters, while Nozaki, et. al. (12) conducted a similar study on methyl p-tolyl sulfoxide.

Thus far the discussion has centered on the observation of chemical shift differences between corresponding nuclei in (R) and (S) enantiomers. Fraser, et. al. (13) have detected an interesting example of nonequivalence for the CH_2 protons of benzyl alcohol, which are enantiotopic by internal comparison (3). In the presence of XI they become diastereotopic and appear as an AB quartet; it was noted that such resolution of enantiotopic nuclei can be used for the determination of geminal coupling constants. Analogous results are observed for the methyl groups of dimethyl sulfoxide ($\Delta\Delta\delta$ = 0.17 ppm) and 2-propanol ($\Delta\Delta\delta$ = 0.04 ppm). Pirkle and co-workers (14) used VIII to demonstrate the meso stereochemistry of the pesticide dieldrin (note the meso

dieldrin

plane bisecting the C_4-C_5 bond). In the presence of the shift reagent, the enantiotopic protons on C_4, C_5, and C_3, C_6 become nonequivalent. The former pair exhibits a vicinal coupling ($J_{4, 5}$ = 3.3 Hz), which establishes the meso geometry. The protons of the corresponding d, l enantiomers are identical in a chiral environment and should thus exhibit no coupling.

The origin of the differential shift, $\Delta\Delta\delta$, has generally been attributed to: (a) differences in stability constants for the diastereomeric adducts shown in eqn. 1, and/or (b) unequal magnitudes for the lanthanide-induced shifts within complexes of equal stability. It appears

that alternative (a) cannot be the sole explanation, since
no apparent correlation exists between the magnitudes
of ΔΔδ and Δδ . Although I and IX cause nearly equal
downfield shifts in benzylmethylsulfoxide, only the latter
can differentiate between the (R) and (S) enantiomers(10).
Similarly, nonequivalence is largest for the least shifted
methyl group of 1, 2-dimethyl-exo-2-norbornanol as shown
in Figure 2c. Perhaps a stronger indication that other
factors are involved is the observation that the sense of
nonequivalence need not be the same for all the nuclei in
a compound. For example, in the presence of I the proton resonances in (R)-1-phenylethylamine are all observed
at lower field than the corresponding protons in the (S)
enantiomer; on the other hand, the methyl protons of the
(R) enantiomer resonate at higher field and the C-H protons at lower field than the analogous (S) protons in the
presence of II or VII (5). Similar behavior is observed
for the protons of 2-phenyl-2-butanol in the presence of
VIII with differing values of the molar shift reagent/sub-
strate ratio. Even more striking is the reversal in the
sense of nonequivalence for the same nucleus. Thus ΔΔδ
for the ortho hydrogens reaches a maximum of ~0. 2 ppm
at a ratio of ~0. 4, decreases to zero at a ratio of ~0. 8,
and is ~0. 3 ppm at a ratio of >2 (9). These results indicate that alternative (b) probably predominates, that
is, spectral nonequivalence results principally from
intrinsically different magnetic environments produced
by the chiral shift reagent.

The identity of the adduct in solution is an important
question which has received relatively little attention to
date. A comparison of the observed dependence of
ΔΔδ on the shift reagent/substrate ratio for a variety
of substrates with that calculated on the basis of a 1:1 or
1:2 adduct led Eikenberry (9) to conclude that for VIII,
a 1:2 adduct is the predominant species. In the case of
IX, a 1:1 adduct is most probable except for amines,
where a 1:2 adduct is again indicated. Deviations from
a perfect fit of calculated and observed behavior were
attributed to possible complicating factors such as self-

association of the shift reagent or the presence of more than one stereoisomer of the tris chelate complex. Some support for the latter suggestion was obtained from a study of the temperature dependence of $\Delta\Delta\delta$ (13). Clearly, more work is needed to clarify the nature of the adducts formed in solution.

Achiral Shift Reagents: Their Use in Distinguishing Diastereotopic Nuclei

It has previously been noted that the association of chiral shift reagents with the (R) and (S) enantiomers of a substrate produces two diastereomeric adducts which, in principle, can be distinguished by their NMR spectra. In contrast, nuclei which are intrinsically diastereotopic are already nonequivalent; achiral shift reagents may then be used to induce larger shift differences. This latter technique is finding increased application in stereochemical investigations and will be briefly discussed in this section.

Green and Shelvin (15) used XII to distinguish the meso and d,l diastereomers of bis(phenylsulfinyl) methane. The CH_2 protons of the racemic pair are equivalent and thus indistinguishable by NMR (unless a chiral shift reagent is used), while those in the meso compound are nonequivalent. Of the two fractions separated by crystallization, the lower melting one exhibits an AB spectrum which identifies it as the meso compound. The higher melting fraction shows a singlet and is thus the racemic mixture.

Wright (16) identified the diastereotopic A and B protons in 1-amphetamine by noting the sensitivity of the induced shifts to the concentration of the added shift reagent tris (6, 6, 7, 7, 8, 8, 8-heptafluoro-2, 2-dimethyl-3, 5 - octanedionato) europium(III), XIII. Differential induced shifts have also been observed for diastereotopic protons in a series of β-lactams (17) and acyclic alcohols (18).

Gerlach and Zagalak (19) reacted primary alcohols with (-)-camphanic acid to form the corresponding esters. The α-methylene protons in the alcohol moiety were observed to be diastereotopic in the presence of XII. Using this approach, the enantiomeric purity of several α-deuterated primary alcohols was determined and an empirical method for assigning absolute configuration proposed.

CONCLUDING REMARKS

The utility of chiral shift reagents should insure their widespread use in the determination of enantiomeric purity. They are effective for a wide range of organic substrates, and normally produce differential shifts large enough to be detected on a 60 MHz instrument (8). One shortcoming of the method to date has been the inability to assign absolute configuration on the basis of similarities in ΔΔδ values. The observation that corresponding nuclei in closely related compounds display reversals in the sense of nonequivalence is not reassuring in this regard (9).

Achiral shift reagents have been shown to magnify the inherent NMR nonequivalence of diastereotopic nuclei. They are thus a valuable tool for making stereochemical assignments, and will undoubtedly receive increasing attention in this role.

REFERENCES

1. W.H. Pirkle and S.D. Beare, J. Amer. Chem. Soc., 91, 5150 (1969), and earlier references cited therein.
2. C.C. Hinckley, J. Amer. Chem. Soc., 91, 5160 (1969); J.K.M. Sanders and D.H. Williams, J. Amer. Chem. Soc., 93, 641 (1971); P.V. Demarco, T.K. Elzey, R.B. Lewis and E. Wenkert, ibid., 92, 5734, 5737 (1970); B.L. Shapiro, J.R. Hlubucek, C.R. Sullivan and L.F. Johnson, ibid., 93, 3281 (1971).
3. M. Raban and K. Mislow, Top. Stereochem., 1, 1 (1967).

4. G. M. Whitesides and D. W. Lewis, J. Amer. Chem. Soc., 92, 6979 (1970).

5. G. M. Whitesides and D. W. Lewis, J. Amer. Chem. Soc., 93, 5914 (1971).

6. R. E. Rondeau and R. E. Sievers, J. Amer. Chem. Soc., 93, 1522 (1971).

7. B. Feibush, M. F. Richardson, R. E. Sievers and C. S. Springer, Jr., J. Amer. Chem. Soc., 94, 6717 (1972).

8. H. L. Goering, J. N. Eikenberry and G. S. Koermer, J. Amer. Chem. Soc., 93, 5913 (1971).

9. J. N. Eikenberry, Ph. D. thesis, University of Wisconsin, Madison, Wisconsin, 1972.

10. R. R. Fraser, M. A. Petit and J. K. Saunders, Chem. Commun., 1450 (1971).

11. E. B. Dongala, A. Solladie-Cavallo, and G. Solladie, Tetrahedron Lett., 4233 (1972).

12. H. Nozaki, K. Yoshino, K. Oshima and Y. Yamamoto, Bull. Chem. Soc. Japan, 45, 3495 (1972).

13. R. R. Fraser, M. A. Petit and M. Miskow, J. Amer. Chem. Soc., 94, 3253 (1972).

14. M. Kainosho, K. Ajisaka, W. Pirkle and S. D. Beare, J. Amer. Chem. Soc., 94, 5924 (1972).

15. J. L. Greene and P. B. Shelvin, Chem. Commun., 1092 (1971).

16. G. E. Wright, Tetrahedron Lett., 1097(1973).

17. A. K. Bose, B. Dayal, H. P. S. Chawla and M. S. Manhas, Tetrahedron Lett., 3599 (1972).

18. P. S. Mariano and R. McElroy, Tetrahedron Lett., 5305 (1972).

19. H. Gerlach and B. Zagalak, Chem. Commun., 274 (1973).

20. V. Schurig, Tetrahedron Lett., 3297 (1972).

21. K. J. Eisentraut and R. E. Sievers, J. Amer. Chem. Soc., 87, 5254 (1965).

Table 1. Several Chiral and Achiral Shift Reagents[a]

Number	Compound	Ref.
(Chiral)		
I	$M = Eu$, $R = C(CH_3)_3$	4
II	$M = Eu$, $R = R_1$	5
III	$M = Eu$, $R = 77\% R_2 + 23\% R_1$	5
IV	$M = Eu$, $R' = R'' = R_1$	5
V	$M = Eu$, $R' = R_1$, $R'' = R_3$	5
VI	$M = Eu$, $R' = 77\% R_2 + 23\% R_1$, $R'' = R_3$	5
VII	$M = Eu$, $R' = R_1$, $R'' = 77\% R_2 + 23\% R_1$	5
VIII	$M = Eu$, $R = CF_3$	8, 9, 20
IX	$M = Eu$, $R = C_3F_7$	9, 10
X	$M = Pr$, $R = CF_3$	7, 9
XI	$M = Pr$, $R = C_3F_7$	9
(Achiral)		
XII	$M = Eu$, $R' = R'' = C(CH_3)_3$	21
XIII	$M = Eu$, $R' = C_3F_7$, $R'' = C(CH_3)_3$	6

[a]VIII-XI, and their Yb(III) analogs, can be purchased from Willow Brook Labs, PO Box 526, Waukesha, Wisc. 53186

Table 2. ΔΔδ Values for Some Chiral Shift Reagents[a]

Compound	Proton	ΔΔδ (ppm) Shift Reagent		
		I	VIII	IX
2-octanol	α-CH$_3$	-[b]	0.11	-
2-methylbutanol	α-CH$_a$	0.00	-	0.05
	α-CH$_b$	0.00	-	0.11
1-phenylethanol	α-H	0.01	0.30	0.07
	CH$_3$	0.00	-	0.05
1,2-dimethyl-endo-2-norbornanol	1-CH$_3$	-	0.37	-
	2-CH$_3$	-	0.33	-
1-phenylethylacetate	-CO$_2$-CCH$_3$	-	0.18	-
1-phenylethylamine	CH$_3$	0.19	-	0.08
	CH	0.41	-	0.00
	H$_{ortho}$	0.12	-	0.06
3,3-dimethyl-2-aminobutane	α-CH$_3$	-	0.28	-
1-methyl-2-norbornanone	1-CH$_3$	-	0.17	-
2-phenylpropanal	α-CH	0.00	-	0.01
	CH(O)	0.00	-	0.03
benzylmethylsulfoxide	CH$_a$	0.00	-	0.06
	CH$_b$	0.00	-	0.03
	CH$_3$	0.00	-	0.06
phenylmethylsulfoxide	CH$_3$	0.09	-	0.11
	H$_{ortho}$	0.02	-	0.12

[a] Spectra taken in CCl$_4$ at molar shift reagent/substrate ratio >0.6; data from refs. 8 and 10.

[b] Indicates data not available.

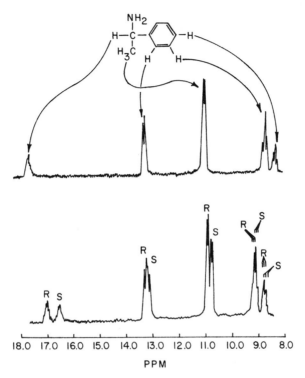

Figure 1.   NMR spectra of (S)-1-phenylethylamine
(upper) and a 7/5 mixture of (R) - and (S)-1-phenyl-
ethylamine (lower), dissolved in CCl_4 in the presence of
I.   The chemical shift scale applies to lower spectrum
only. (Taken from ref. 4, courtesy of the American
Chemical Society).

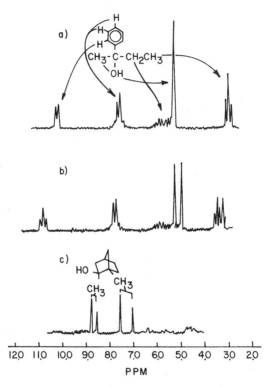

Figure 2. NMR spectra of 2-phenyl-2-butanol dissolved in CCl$_4$ in the presence of (a) XII and (b) VIII, and (C) the spectrum of 1,2-dimethyl-exo-2-norbornanol in the presence of VIII. (Taken from ref. 8, courtesy of the American Chemical Society).

STRUCTURE ELUCIDATION OF NATURAL PRODUCTS[1a]

R. Burton Lewis[1b-d] and Ernest Wenkert
Department of Chemistry, Indiana University
Bloomington, Indiana   47401

Two methods for the use of Lanthanide Shift Agents
have evolved: the qualitative approach using the coupling
constants and approximately extrapolatable original chemical shifts from a shifted spectrum, and allowing structure assignment on the basis of available empirical
knowledge; and the quantitative approach placing heavy
dependence upon the magnitudes and directions of the
shifts for the calculation of the three-dimensional
structures.[2] The calculations, generally computerized,
use various correlation coefficients to measure the
correspondence between the computed and observed values.
Our approach was semi-quantitative for several reasons,
mainly because in most structure problems the basic
molecular skeleton is known from the nature of the
starting materials or the methods of isolation.   Therefore the problem is often locating the site or orientation of functionality.   The present work examined the
potential of these agents for spectral simplification,
location of functionality and the parameters affecting
the magnitude of the observed shifts.

Characterization of the "ideal" shift agent organizes
the points important to experimental work.   The rank of
importance of these criteria varies according to the
specific application involved, not all criteria being
mutually exclusive. (1) Most importantly, the shifted
resonances should remain sharp for all compounds. (2) The
coupling constants which would be measured in the shifted
spectrum should be truly constant.[3] (3) The induced shifts
should be large and predictable (pseudocontact). (4) There
should be strong differentiation among functional groups,
particularly those within one compound.   The shifts should
be (5) reasonably solvent-independent, (6) qualitatively
temperature-independent, and (7) concentration independ-

ent/ratio dependent. (8) Further, the agent and compound should be recoverable and (9) the enantiomer ratios available, if desired. Many of these conditions hinge upon one consideration: there should be a single complex in solution which is not dependent upon the functional group, concentration or steric bulk of the substrate ligands.

The effect of the alteration of the structural features of one functional type was investigated, allowing an assessment of several variables. One of these is the effect of steric factors on coordination numbers and binding constants,[5,6] and rates of exchange of bound with free substrate. A second factor, the magnitude of the variation of the Knight shift with distance and angle, gives an insight into the dominant mechanism of action of the shift agent. Thirdly, the ligation of the functional group to the Eu(dpm)$_3$ in the case of non-rigid ligands might cause changes in the effective steric size of that functional group or in the electronic factors in the Lewis base resulting in changes in the populations of conformers or in shifts not due to the paramagnetism of the metal.

To begin to gain insight into some of these points, the four isomers of butanol were used as ligands and the chemical shifts recorded as a function of added Eu(dpm)$_3$. The spectra were obtained at relatively high alcohol concentration and low Eu(dpm)$_3$ concentration in order to attempt to eliminate changes in the coordination number as one of the possible effects, allowing less ambiguous assessment of other factors involved. The quantitative expression of the magnitude of shift is given in ppm as Δ_{Eu} value.[7]

The butanol spectra show all shifts to be isotropic, i.e., completely averaged, and remain relatively sharp through all concentrations of Eu(dpm)$_3$, as noted by others.[8,9,10,11] Thus bulk of alcohols does not affect significantly the rate of exchange on the nmr time scale. Further, all shifts are downfield, large and qualitatively a simple function of the distance from the functional group (Figure 1).[8,9,11,12] This results in very similar values for similarly positioned protons. For example, the variation of the value for the carbinol methine proton, -24.50, -26.40, -26.00, is less than two ppm or ca. 4% maximum deviation from the average of -25.63 ppm. The beta and gamma protons show similarly close figures. These data imply that the nature of the Knight shift is

100

predominantly pseudocontact.[8,9,10,11,12,13] As there are
no changes in the sign of the shift, these small compounds make only small deviations in angle from the magnetic axis. This evaluation being so straightforward
must derive from the steric bulk of these ligands not
markedly affecting the equilibrium constant for the association of the alcohol with the metal, at least for those
conformations of the complex which lead to a shift.

   Another method for the evaluation of the similarity
of effect is a ratio calculation. The ratio of all
proton values to the value of the carbinol methine proton, for example, eliminates the effect of the g factors,
the measure of the shifting ability of the metal. This
leaves a geometric ratio which is dependent only upon
angles and distances of the protons from the metal. The
ratio data for the butanols (Figure 1) allows a qualified
judgement to be made about the predominant conformational
isomers of the population leading to the shifts. Since
a particular conformer has a characteristic set of Δ_{Eu}
values due to its particular set of angles and distances,
the conformer population as a whole has a resultant value
dependent upon the weighted averages. One major qualification in the evaluation is the assumption that the
various conformers bind to the europium in the same proportions in which they exist in solution. If this is
incorrect, there is over-representation of the contribution of some isomer(s). Since steric size does not seem
to make a large difference in the absolute magnitude of
the shifts, the assumption seems reasonable. The hydroxyl
values remaining amazingly invariant indicates there to
be relatively little variation in the behavior and location of the hydrogen from one compound to the next. The
hydroxyl hydrogen for <u>tert</u>-butyl alcohol is defined to
be 3.71 and the ratio for the beta protons calculated
accordingly to be 0.605. While the Eu complex probably
does not bind in a Eu-O-C linear array, rotation and/or
exchange around the O-Eu bond probably results in averaging to such a position. Hence in <u>n</u>-butyl alcohol,
europium is tilted slightly away from the beta methylene
group, as the value for the beta protons is low, 0.567.
In the case of <u>sec</u>-butyl alcohol the europium seems to be
shifted away from the α-methyl and toward the α-methylene
unit, shown by the low value for the methyl group, 0.554,
and the higher values for the methylene protons, 0.575

and 0.668. Further, the magnetic non-equivalence of the methylene group is accentuated by the shift effect of the europium. The shifted spectra of 3-pentanol also show individual resonances for the β-methylene protons, corresponding closely with the ones for sec-butyl alcohol. In order to determine which of the methylene protons was most shifted, a pmr study of erythro-3-deuterio-2-butanol was undertaken. The shifted spectra of the alcohol show five of the usual six resonances. The Δ_{Eu} values and ratios were slightly different from those of 2-butanol, but clearly consistent with the above assignments.

Two nuclei are said to be equivalent if the sums of all the local magnetic fields due to surrounding nuclei yield the same value. The local magnetic fields for the diastereomeric β-protons do not average to the same value. Being very close, however, they are not seen as separate signals in the diamagnetic spectrum, but only broaden the normally expected signal. The effect of the Eu complex is to accentuate the magnetic non-equivalence. A further contribution to the non-equivalence may be conformational preferences.

An example of conformational preference involving only the Eu-O bond is 1-(1-hydroxyethyl)- adamantane. In this compound there can be no conformational preference about the adamantane-ethanol bond and yet there is a significant difference in the Δ_{Eu} values between the γ-diasterotopic protons, absent in the spectrum without the Eu complex. The origin of the shift difference may reside in the Eu complex binding in a fashion other than along the C-O bond extension.[9] A number of other simple alcohols are listed in Table 1 to show the applicability of the shift agent. Phenol, giving severely broadened spectra and low Δ_{Eu} values, will be discussed later.

In order to find the differences resulting from axial and equatorial hydroxyl groups on a cyclohexane ring, the shifted spectra of cis- and trans- to 4-t-butyl cyclohexanol were obtained.[7] The shifted spectra for the cis alcohol, seen in Figure 2, show every type of proton as a separate resonance, each assigned on the basis of multiplicity and decoupling experiments. The graphical presentation of these data shows that the resonance positions of the protons are divergent. This is not the situation for the trans alcohol. In this case there is an unfortunate combination of initial chemical shift dif-

102

ferences and Δ_{Eu} values leading to two sets of overlapping resonances for five proton positions.

The shifted spectra of two compounds, 2-methyl-2-adamantanol and 1-adamantanol, can be evaluated by the use of the t-butylcyclohexanols as models. Their data (Table 2) show a close correlation, especially considering the steric changes in the compounds, when compared by the ratio method the ratios for each proton are derived from its equatorial γ-proton.

Demarco and Elzey[14] applied these results to 2-β-androstanol (Figure 3). The unshifted spectrum (A) shows little information, while the 100 MHz spectrum (B) shows all of the ring A protons upon addition of one mole equivalent of Eu(dpm)₃. They shift in the order of 2α, equatorial 1β and 3β, axial 4β, axial 1α and 3α, equatorial 4α and, finally, axial 5α, just as predicted from cis-4-t-butylcyclohexanol. The 220 MHz spectrum (C) reveals most of the protons in ring B and some in ring C. The analysis of 3-β-friedelanol[15] (Figure 4) follows substantially the same pattern, although the ordering is not completely followed in this case and possibly in the previous case with regard to the equatorial and axial protons γ and δ, respectively (to the hydroxyl group). Since the diamagnetic position of axial protons is generally upfield to that of equatorial hydrogens, the axial δ protons seem ·to be shifting more strongly than the equatorial γ ones, indicating perhaps that the model compound, cis-4-t-butylcyclohexanol, is slightly distorted with respect to the geometries found in normal steroid and triterpene ring systems. Some other natural products which have been examined are lupeol,[11] cholesterol monohydrate,[10] testosterone and 17-α-methyltestosterone,[12] triterpenoids,[16] trachyloban-19-ol,[17] and amino steroids[18] Still other compounds are borneol and iso-borneol,[7] 6-deuterio-borneol,[19] 2-norbornanols and dehydronor-borneols,[20] exo-norbornenol and nortricyclenone,[21] and exo-3-chloro-endo-5-acetoxy- and exo-3-chloro-exo-5-acetoxynortricyclene.[22] Further ring compounds include cyclopropylcarbinols,[23] azetidine alcohols,[24] trimethyl-cyclohexanols[25] and perhydrophenalenols.[26]

As no two Lewis bases interact with a Lewis acid in exactly a like manner, differences can be expected on changing the functional group within the same structural framework. For the determination of these differences the

four butyl amines were examined for direct comparison with
the butyl alcohols. Furthermore, a wide variety of functional groups of varying structural features was investigated.

Butyl amines, run at high concentrations and low
Eu(dpm)$_3$ ratios, behave qualitatively like the alcohols.
Thus, line broadening was minimal and the shifts were
large and in approximate proportion to the distance.
Other work has shown that Yb(dpm)$_3$ shifts of amines can
be very large.[27] The plots of the shifts <u>versus</u> concentration of Eu(dpm)$_3$ yielded straight lines from which
the Δ_{Eu} values were determined (Figure 5). These values
for α-protons are uniformly 33% larger than corresponding
values for the alcohols. On the assumption of the ratio
for the amino hydrogens, observed in only one case, being
the same as for hydroxyl hydrogen, the approximate value
for the <u>tert</u>-butyl protons, 0.708, seems reasonable. The
ratios for the β- and γ- protons for <u>n</u>-butyl amine, 0.622
and 0.363, do not match exactly those for <u>n</u>-butyl alcohol,
0.567 and 0.396; while the δ-proton values are identical,
0.188. The ratio for the beta protons being larger than
that for the same protons in the alcohol, 0.622 vs. 0.567,
implies that these protons are closer to the metal atom
in the amine complex. The γ-protons, in contrast, show
a ratio which is smaller than that for the alcohol, 0.363
vs. 0.396, consistent with the idea that these protons are
at a distance farther from the Eu complex. This implies
that there is a larger proportion of <u>trans</u> conformation
for the amine than for the alcohol. (This analysis
depends upon the contact contribution to the Knight shift
for protons remaining small. Otherwise, the contact contribution must be corrected and the data reanalyzed.) <u>Sec</u>-
butyl amine shows the same accentuation of the magnetic
non-equivalence of the diastereotopic methylene protons
as <u>sec</u>-butyl alcohol. Since geometric ratios for <u>sec</u>-
butyl amine protons mirror those of 2-butanol, it seems
the predominant conformer of both is that shown.

The survey of the functional groups (Table 3), exclusive of the alcohols, shows several important trends.
The primary observation, hard bases tending to be more
strongly shifted than soft ones, is in consonance with the
fact that the rare earths are hard acids and therefore
bind hard bases best,[28] i.e., F>O>N>C and O>S>Se, in order
of decreasing hardness. In view of this fact a reason

other than stronger binding must be found to explain the
increased shifts of amines relative to alcohols. The
hardness of a base decreases as its electrons are delocalized in any fashion, e.g., a compound such as nitro-
propane (13) shows almost no interaction with the Eu-
(dpm)$_3$. Conversely, an ether such as tetrahydrofuran (1),
a strong base, shows large shifts. Butyl ether (2), not
as strong a base as THF and more bulky, shows a much decreased shift. Anisole (3), having its ether electrons
delocalized in the benzene ring, shows even less shift.
Comparisons of pyrrolidine (23), tetrahydrofuran and
tetramethylene sulfide (30), as well as tetramethylene
sulfide, tetramethylene sulfoxide (14) and tetramethylene
sulfone (13), also show clearly the difference between
hard and soft bases. Thus the primary factors of binding
to Eu(dpm)$_3$ are the hardness of the base, its acidity[29]
and its steric bulk.

Solvent and concentration effects are so large as to
necessitate exact matching of experimental conditions for
comparison of the magnitudes of shifts among various
compounds. For example, the plots of the shifted spectra
of 2-butanol in three solvents, cyclohexane, CCl$_4$ and
CHCl$_3$ show that the shift per increment of Eu(dpm)$_3$ falls
to nearly zero at higher concentrations of Eu(dpm)$_3$. The
Δ_{Eu} values are different, and are consistently largest
for cyclohexane and smallest for chloroform. Plots of
the ratios of the shifts for all Eu(dpm)$_3$ concentrations
show that the geometric ratios appear to be the same for
any one solvent, although the slopes change. This means
that the basic form of the complex and the orientation
and magnitude of the magnetic vector of the metal are
similar. Substrate concentration also has a profound
effect on the actual shift. For example, a series of
spectra which have identical alcohol/Eu(dpm)$_3$ ratios, but
differ in concentration, differ in shifts, decreasing
markedly at lower concentrations. Thus, Eu(dpm)$_3$ does not
form a discrete complex with the ligand which exchanges
with excess ligand, but establishes a steady state concentration of the adduct species which is dependent upon
the concentration of both Eu(dpm)$_3$ and the alcohol.

As adduct formation is an equilibrium process,
temperature plays a role in the determination of the final
concentrations of the organometallic species involved.
Lower temperature favors greater equilibrium concentra-

tions of species which lead to downfield shifts in resonance positions of n-hexyl amine protons. A second effect
of temperature on shifts shows the geometric ratios
changing on alteration of the temperature. More subtle
effects in the complex are evident when the changes in a
sterically rigid ligand such as cis- or trans-4-t-butyl-
cyclohexanol are examined. These shifts (i.e., Figure 6)
indicate that the magnetic vector (and thus metal atom)
changes position and orientation with respect to the
alcohol ligand. This is possible only if more than one
species in solution is responsible for the shifts and
their relative populations change with temperature.

This thesis is supported by the observation of unshifted resonances attributable to free dipivaloylmethane
at lower temperatures, all protons, the methyls, methine
and hydroxyl, being visible. Apparently alkoxide complexes are being formed.

Formulation of the interactions occurring upon
addition of Eu(dpm)$_3$ to an alcohol solution is now possible, given several assumptions. Firstly, the contribution
of species which have coordination numbers of six or less
is negligible, entirely consonant with the structure data
from x-ray analyses[30,31,32,33,34,46,47] and the chemistry
of rare earth compounds.[30,35,36,37] Secondly, the coordination number is not more than eight. Since the Eu-
(dpm)$_3$ complex is even bulkier than most normal complexes
and coordination numbers of more than eight have been
demonstrated definitely only for salt complexes, such
coordination numbers should not be evident here. Thirdly,
the formation of dimer complexes containing more than one
alkoxide radical per Eu atom is assumed to be less favorable than addition of dpmH to that complex.[38] Formulation
of structures analogous to those demonstrated by x-ray
determinations leads to the structure chart shown in
Figure 7. A wide variety of monomer and dimer forms may
contribute to the total population of the complex, thereby giving a rationale for the liberation of dpmH under
certain circumstances. Also changes in the geometric
ratios result from shifting equilibrium concentrations of
the species due to changes in temperature or reactant concentration.

A scheme involving dimers as well as monomers
accounts for the failure of Eu(dpm)$_3$ to give readily
interpretable spectra when applied to some compounds.

Phenol gives shifted spectra broadened more than those for tertiary alcohols and having smaller Δ_{Eu} values than expected from a model like 1-adamantanol. The spectra show a large peak in the region for free dpmH, indicating that phenol, a stronger acid than the other alcohols, forms higher concentrations of bridged chelate dimers (i.e., V or VI). Since this results in species in which the phenolate is more strongly bonded, the exchange rate is slowed and the resonances broadened. The lower shifts for phenol may be the result of smaller g factors or changed orientation of the magnetic axes for these complexes. As the x-ray analysis of bridged dimers has shown the bridging distances to be longer than those non-bridging,[30] this may be another reason for the smaller shifts.

Other compounds whose spectra gave broadened signals were 2-ethoxyethanol and 1,3-propanediol. As a number of dimer complexes (i.e., III, IV, V, VI) and a monomer complex (VIII) offer the possibility of strong bidentate binding of the ligand molecule, slowed exchange leading to broadened lines may result. For these compounds the broadening can result in widths of nearly one ppm at half height. Other compounds showing significant broadening are monosaccharides.[39,40,41] 2-Hydroxymethylfuran does not have a strong binding site at the ether function, resulting in less broadened lines and more normal shifts.

On the basis of the proposed complexes a reason for amine shifts larger than those of alcohols may be postulated. It was observed during the temperature dependence work that the quantity of free dpmH was much smaller for the amine than for the alcohol. This implies that nitrogen is not as good a bridging species as oxygen, as would be expected on the basis of the former being a softer base. As bridging species seem not to have shifts as large as monomer species based upon observations with phenol, it may be hypothesized that oxygen bases bind more strongly than amines but also form higher concentrations of bridged species, thereby leading to lower net shifts.

A problem of structural interest involved the natural product piperine,[42] whose three functional groups could lead to unusual spectra. The compound could not be expected to ligate in a bidentate fashion. The results shown in Figures 8 and 9 and Table 4, are interesting for several reasons. Most importantly, the ether oxygens do

not complex in competition with the amide group as indicated by lack of shift in the methylenedioxy group up to
an equimolar ratio of piperine to Eu(dpm)$_3$. Thus, easily
interpretable spectra are obtained. Secondly, the angle
variables are extremely important in this compound as the
γ- and δ- vinyl protons as well as all the protons of the
benzene ring are shifted upfield. This indicates that
their positions or their averaged positions lie outside
the 54° line of the magnetic axis of the metal. This same
effect has been observed by others.[43,44,45] The piperidine unit is seen as averaged in the diamagnetic sample,
i.e., the rotation about the N-carbonyl carbon bond is fast
on the nmr time scale. Addition of Eu(dpm)$_3$ in either
carbon tetrachloride or chloroform solutions results in
spectra which show individual resonances for the α-, β-
and γ- methylene groups. Separation into individual
resonances occurs sooner in CHCl$_3$, probably due to the
ability to hydrogen bond to the carbonyl oxygen, thereby
raising the energy barrier to rotation. The primary
reason for the ability to see the individual resonances
could evolve simply from the greater difference in chemical shift between the two positions, induced by the
Eu(dpm)$_3$, or by the action of the complex as a Lewis acid
serving to polarize the molecule and increase the double
bond character of the N-C bond. The interaction of
La(dpm)$_3$, a diamagnetic rare earth complex, with piperine
in chloroform leads to a separation of the two alpha
methylene signals, but only near an equimolar ratio of
compound and europium complex. This seems to indicate
that the acid-base polarization of piperine contributes,
but is not the determining feature, in obtaining individual resonances. It shows further that the increase in the
chemical shift difference is the predominant factor. The
La study also shows the contributions to the induced shift
due to factors other than the paramagnetism of the Eu ion
are negligible, -0.1 ppm or less.

The spectra reveal a second phenomenon which is
superfically the same. The peak corresponding to the β-
vinyl proton is a doublet of doublets of doublets in both
chloroform and carbon tetrachloride solutions with couplings of 14, 6 and 4 Hz and of 14, 8 and 3 Hz, respectively. Comparison of spectra of the chloroform solution
taken at 60, 100 and 220 MHz confirmed this as coupling
corresponding perhaps to two conformations of the piperine

108

and not two superimposed signals. As La(dpm)$_3$ or Eu(dpm)$_3$
is added to the solution, signals for both α- and β- vinyl
protons broaden and, as an equimolar ratio is approached,
again sharpen. The α- proton is a doublet and the coupling is unchanged throughout the course of the addition.
The β- proton is revealed upon sharpening to be only a
doublet of doublets with coupling of 14 and 11 Hz in
either solvent. Separate resonances corresponding to more
than one conformation of the piperine are not observed for
either of these protons. There seem to be several possible explanations for these phenomena. The γ- and δ-
protons do not show any broadening or change in coupling
patterns from the first increment of added Eu(dpm)$_3$ on-
ward, but are not observable in the diamagnetic spectrum.
Inasmuch as no proton in the spectrum seems to have
additional coupling after addition of 10 mole per cent of
Eu(dpm)$_3$, it seems that the third coupling in the beta
proton has disappeared by this time. This indicates that
its broadening and change in the coupling pattern are not
related. Thus, the broadening in the α- and β- signals
must be due to slowed rotation about the carbonyl α-carbon
bond resulting in a single conformation, as no additional
signals for these protons are observed. The coupling
observed in the β- proton may be due to a γ-, δ- or phenyl
proton in a conformer wherein the two vinyl groups are not
coplanar. Addition of small amounts of rare earth complex
is enough to cause coplanarity of the natural product and
consequent disappearance of the coupling.

The figures indicate that, while the chloroform and
carbon tetrachloride solutions are qualitatively the same,
all shifts in both solvents being of corresponding sign,
there are some quantitative differences. For example,
the magnitude of the observed shifts are higher for the
carbon tetrachloride solutions, indicating higher equilibrium constants for formation of the adducts in this
solvent as in the case for 2-butanol. Furthermore, the
ratios for the chloroform solutions are larger than those
for the carbon tetrachloride solutions, implying a longer
equilibrium bond distance in chloroform. The breaks in
the curves indicate that in carbon tetrachloride more
selective complexation takes place, perhaps due to the
competition of CHCl$_3$ with Eu(dpm)$_3$ for the amide group.

Finally, there are different values for Ar(2) and
Ar(6) indicating that the benzene ring does not rotate

with complete freedom and must have a preferred conformation. Each conformation of the benzene ring has a
unique set of Δ_{Eu} values, the observed value being determined by the relative populations of each. Models indicate that each aromatic proton in one of the conformers
has upfield contributions and in the other conformation
downfield contributions, with the former being the larger.
This leads to the conformation shown in the figure as the
predominant contribution.

Footnotes

1. (a) This work taken in part from a dissertation presented by RBL in partial fulfillment of the requirements for the Ph.D. degree, Indiana University, 1971. (b) U. S. Public Health Service, Fellow, 1969-1971. (c) Author to whom correspondence should be addressed. (d) Present address: Western Electric, Box 900, Princeton, N.J., 08540.

2. For several fine examples of quantitative work, see other chapters of this book.

3. Small changes are found.[4]

4. B.L. Shapiro, M.D. Johnston, Jr., R.L.R. Towns, J. Am. Chem. Soc., 94, 4381 (1971).

5. K. Nakamura, J. Inorg. Nucl. Chem., 31, 455 (1969).

6. K. Nakamura, J. Inorg. Nucl. Chem., 32, 2265 (1970).

7. P.V. Demarco, T.K. Elzey, R.B. Lewis and E. Wenkert, J. Amer. Chem. Soc., 92, 5734 (1970).

8. G.H. Wahl, Jr. and M.R. Peterson, Jr., Chem. Comm., 1167 (1970).

9. J.K.M. Sanders and D.H. Williams, J. Am. Chem. Soc., 93, 641 (1971).

10. C.C. Hinckley, J. Am. Chem. Soc., 91, 5160 (1969).

11. J.K.M. Sanders and D.H. Williams, Chem. Comm., 422 (1970).

12. C.C. Hinckley, R.M. Kloty and F. Patil, J. Am. Chem. Soc., 93, 2417 (1971).

13. A. Carrington and A.D. McLachlan, "Introduction to Magnetic Resonance," Harper and Row Publishers, New York, 1967, page 221.

14. P.V. Demarco, T.K. Elzey, R.B. Lewis and E. Wenkert, J. Amer. Chem. Soc., 92, 5737 (1970).

15. N.K. Dutt, S. Rahut and S. Sur, J. Inorg. Nucl. Chem., 33, 121-125 (1971).

16. D.G. Buckley, G.H. Green, E. Ritchie and W.C. Taylor Chem. Ind., 298 (1971).

17. O. Achmatowicz, Jr., A. Ejchart, J. Jurczak, L. Kozersaki and J. St. Pyrek, Chem. Comm., 98 (1971).

18. L. Lacombe, F. Khuong-Huu, A. Pancrazi, Q. Khuong-Huu and G. Lukacs, Comptes Rendus Hebdomadaires des Seances (C), 272, 668 (1971).

19. K. Tori, Y. Yoshimura and R. Muneyuki, Tet. Lett.,

333 (1971).

20. J. Paasivirta, Suomen Kemistilehti (B), 44, 131 (1971).

21. E. Vedejs and M.F. Salomon, J. Am. Chem. Soc., 92, 6965 (1970).

22. G.M. Whitesides and J. San Filippo, Jr., ibid., 92, 6611 (1970).

23. M. Yoshimoto, T. Hiraoka, H. Kuwano and Y. Kishida, Chemical and Pharmaceutical Bull. (Japan), 19, 849 (1971).

24. T. Okutani, A. Morimoto, T. Kaneko and K. Masuda, Tet. Lett., 1115 (1971).

25. P. Belanger, C. Freppel, D. Tizane and J.C. Richer, Can. J. Chem., 49, 1985 (1971).

26. F.A. Carey, J. Org. Chem., 36, 2199 (1971).

27. C. Beaute, Z.W. Wolkowski and N. Thoai, Tet. Lett., 817 (1971).

28. R.G. Pearson, J. Chem. Educ., 45, 581, 643 (1968).

29. L. Ernst and A. Mannschreck, Tetrahedron Letters, 3023 (1971).

30. C.S. Erasmus and J.C.A. Boeyens, Acta Crystallographica, A25, S3, p. S162, (1969).

31. A. Zalkin, D.H. Templeton and D.G. Karraker, Inorg. Chem., 8, 2680 (1969).

32. M.F. Richardson, P.W.R. Corfield, D.E. Sands and R.E. Sievers, Inorg. Chem., 9, 1632 (1970).

33. J.A. Cunningham, D.E. Sands, W.F. Wagner and M.F. Richardson, Inorg. Chem., 8, 22 (1969).

34. F.A. Cotton and P. Legzdins, ibid., 7, 1777 (1968).

35. N. Filipescu, C.R. Hurt and N. McAvoy, J. Inorg. Nucl. Chem., 28, 1753 (1966).

36. M. Hasan, S.N. Misra and R.N. Kapoor, Indian J. Chem., 7, 519 (1969).

37. B.S. Sankhla and R.N. Kapoor, Can. J. Chem., 44, 1369 (1966).

38. T.R. Sweet and H.W. Parlett, Analytical Chemistry, 40, 1885 (1968).

39. R.F. Butterworth, A.G. Pernet, and S. Hanessian, Can. J. Chem., 49, 981 (1971).

40. I. Armitage and L.D. Hall, Chem. Ind., 1537 (1970).

41. P. Girard, H. Kagan and S. David, Soc. Chimique De France Bull., 4515 (1970).

42. E. Wenkert, D.W. Cochran, E.W. Hagaman, R.B. Lewis and F.M. Schell, J. Amer. Chem. Soc., 93, 6271

(1971).

43. M. R. Willcott, J.F.M. Oth, J. Thio, G. Plinke and G. Schroder, Tet. Lett., 1579 (1971).

44. P. H. Mazzochi, H.J. Tamburin and G. R. Miller, ibid., 1819 (1971).

45. T.H. Siddall, III, ibid., 452 (1971).

46. W. De W. Horrocks, J.P. Sipe, III, and J.R. Luber, J. Amer. Chem. Soc., 93, 5258 (1971).

47. J.J. Uebel and R.M. Wing, ibid., 94, 8910 (1972).

Compound	OH	H_{α}	H_{β}	H_{γ}	H_{δ}	H_{ε}	H_{ω}
CH_3OH	100.	24.00					
1-butanol	90.60	24.50	13.90	9.70	4.60		
1-decanol		23.50	13.60	9.20	4.60	2.70	0.50
iso-butanol	98.40	26.40	15.75	9.80			
tert-butanol	98.20		16.00				
1-adamantanol	98.20		17.70	5.40	(cis) 5.85	(trans) 4.60	
allyl alcohol	93.20	24.20	12.70	12.00(cis)	6.70(trans)		
phenol			12.60	4.00	3.10		
benzyl alcohol	95.0	24.90		11.60	3.80	3.00	
$C_6H_5CH_2OH$	88.60	23.40	12.70		6.80	2.80	1.90
$C_6H_5CH_2CH_2OH$		23.70	13.40	9.80		4.20	1.40–0.70
sec-butanol	97.40	26.00	14.40	erythro 16.10, threo 14.40	8.90		
3-pentanol	93.80	24.30	13.40	erythro 14.60, threo 12.70	8.60		
1-(1-hydroxy-ethyl)- adamantane		25.12	13.40	(11.32, 10.48)	3.32	(cis) 3.68	(trans) 2.76
sec-butanol-d₁		25.00	14.15	15.60	8.35		

Table 1. Eu(dpm)₃ shifts, in ppm downfield. Spectra taken on a Varian HA-100 spectrometer at 32°C, at 0.675M solution in CHCl₃ or CDCl₃.

114

Compound		α	ρ	γ	δ	ω
cis-4-t-butyl- cyclohexanol	eq.	24.7(3.69)	14.9(2.22)	6.7(1.00)		2.8(0.42)
	ax.		8.2(1.22)	13.6(2.03)	6.5(0.97)	
trans-4-t-butyl- cyclohexanol	eq.		13.6(2.96)	4.4(1.00)		1.8(0.39)
	ax.	21.7(4.72)	14.7(3.09)	5.4(1.17)	5.4(1.17)	
1-adamantanol	eq.		17.2(3.31)	5.2(1.00)	4.2(0.81)	
	ax.		17.2(3.31)		6.0(1.15)	
A	eq.	(CH₃) 14.72(2.05)	15.6(2.17)	7.2(1.00)	4.5(0.63)	4.1(0.57)
	ax.			16.0(2.22)		
B Ring	eq.	(CH₃) 14.72(2.83)	15.6(3.00)	5.2(1.00)	4.3(0.83)	4.1(0.79)
	Ring ax.			6.4(1.23)		

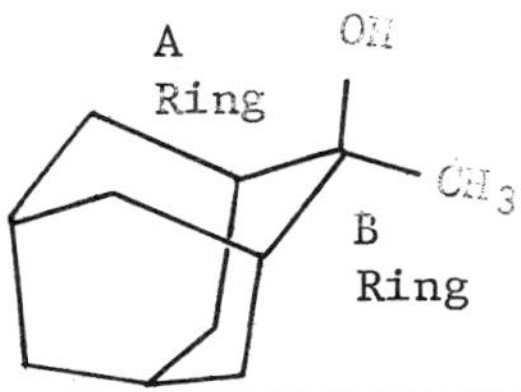

A OH Ring CH₃ B

Table 2. Eu(dpm)₃ shifts, in ppm downfield (ratios in parentheses). Spectra taken on a Varian HA-100 spectrometer at 32°C, 0.32 M solution in CHCl₃ or CDCl₃.

115

	Compound[a]	X–H	H_α	H_β	H_γ	H_δ	H_ε	H_ω
oxygen bases	1. Tetrahydrofuran[b,c]		28.80	13.00				
	2. $(\underline{n}\text{-Bu})_2\text{O}$[c,d]		4.85	3.00	1.70	0.40		
	3. Anisole[b,e]		2.30	(est.)				
oxygen bases	4. n-Butyraldehyde[c,d]		18.50	10.60	7.30	4.15		
	5. $CH_3CH_2CH_2O\overset{O}{C}CH_3(\alpha)$[d]		10.10	10.30	7.25	3.90		
	6. Cyclopentanone[b,c]		9.44	3.56				
esters	7. $CH_3CH_2CH_2CH_2O\overset{O}{C}CH_3(\alpha)$[c,d]		12.30	13.60	5.05	3.45	1.70	
amides	8. $C_6H_5HN\overset{O}{C}CH_3(\alpha)$[b,e]	10.48	10.56	9.56	1.48	1.24		
	9. Pyrrolidone[b,c]		12.88	4.82	4.04			
	10. N-Methyl$^\varepsilon$ pyrrolidone[b,c]		11.48	4.92	5.64	9.24		
	11. (structure)	13.84	16.40	4.04	1.72	1.12	0.44	
	12. (structure)		18.88	6.00	6.60	9.00	21.56	
oxygen attached to hetero atoms	13. Nitropropane[c,d]		0.85	0.85	0.35	(est.)		
	14. Tetramethylene sulfoxide[b,c] cis		10.44	7.24				
	trans		5.36	4.64				
	15. Tetramethylene sulfone[b,c]		6.36	2.64				

Compound	X-H	H_α	H_β	H_γ	H_δ	H_ϵ	H_ω
16. (n-Bu)$_3$P=O [b,c]		7.56	4.84	1.64	0.48		
Nitrogen bases							
17. n-Butyl amine [c,d]		32.50	20.20	11.80	6.10		
18. iso-Butyl amine [c,d]		35.20	20.20	11.80			
19. n-Hexyl amine [e,f,g]		29.64	18.96	10.40	5.64	3.46	2.04
20. CH$_3$$\beta$-CHCH$_2CH_3$ ϵ (NH$_2$ on α) [c,d]		33.40	24.40	22.30	19.20	11.10	
21. Cyclohexyl amine [b,c]		26.30					
22. tert-Butyl amine [c,d]	93.90		17.90				
23. Pyrrolidine [b,c]		37.68	16.64				
24. Piperidine [b,c]		38.20	15.60	13.68			
25. N-Methyl$^\alpha$ pyrrolidine [c,f]		20.52	23.56	10.84			
26. N-Methyl$^\alpha$ piperidine [b,c]		16.00	18.40	10.56	8.00		
27. Pyridine [b,c] (γ α β δ)		26.30	8.84	8.16			
28. CH$_3$CH$_2$C(NC$_6$H$_5$)CH$_2$CH$_3$		6.72	5.56	2.96	2.00		
29. Propylcyanide [c,d]		7.25	3.90	2.85	(est.)		
Sulfur							
30. Tetramethylene sulfide [b,c]		1.60	1.04 (est.)				

Table 3. Eu(dpm)$_3$ shifts, in ppm, downfield. a Spectra taken on a Varian HA-100 spectrometer at 32°C. b 0.32 M solution. c CHCl$_3$. d 0.675 M solution. e CDCl$_3$. f 0.20 M solution. g 29°C.

117

	δ (in ppm)[d] CHCl₃[a,b]				δ (in ppm)[d]		CCl₄[a,c]	Ratios	
	Obsd[f]	Extrap[g]	Δ_La[e]	Δ_Eu	Obsd[f]	Extrap[g]	Δ_Eu[e]	CCl₄	CHCl₃
Ar(C₂)	6.92	6.93	0	+0.10	6.87	6.88	+0.12	+.009	+.009
Ar(C₅)	6.72	6.71	0	+0.27	h	6.63	+0.32	+.024	+.024
Ar(C₆)	6.82	6.81	b	+0.77	6.81	6.83	+0.86	+.063	+.067
αC	6.38	6.38	+0.08	−8.32	6.29	6.26	−9.26	−.684	−.727
βC	7.38	7.38	−0.17	−9.43	7.26	7.26	−9.54	−.705	−.824
γC	h	6.80	>0	−2.24	h	6.69	−2.39	−.177	−.196
δC	h	6.73	<0	+2.21	h	6.64	+2.50	+.185	+.193
Pip(C₂)	3.49[i]	3.49	−0.14[g]	−11.43	3.50	3.50	−13.53	1.000	1.000
Pip(C₃)	1.62	1.70	~0	−3.36	1.61	1.58	−3.84	−.284	−.293
Pip(C₄)	1.62	1.56	~0	−2.14	1.61	1.66	−2.28	−.168	−.187
Pip(C₅)	1.62	1.70	~0	−2.70	1.61	1.58	−3.44	−.254	−.236
Pip(C₆)	3.59[i]	3.56	+0.06[j]	−5.73	3.50	3.60	−6.42	−.474	−.501
Dioxy CH₂	5.92	5.92	0	−0.09	5.92	5.89	−0.09	−.007	−.008

Table 4. [a] Spectra taken on a Varian HA-100 spectrometer. [b] 0.25 M solution at 32°. [c] 0.20 M solution at 32°. [d] In parts per million downfield from internal TMS. [e] $\Delta_M = \delta_1 - \delta$ complex, where complex= 1:1 M(dpm)₃/~; + Δ_M signify upfield shifts. [f] Observed values. [g] Extrapolated values (see Fig. 8 & 9). [h] Overlapping multiplets prevent measurement. [i] Determined at 220 MHz and 13° (below the coalescence temperature). [j] Values may be reversed.

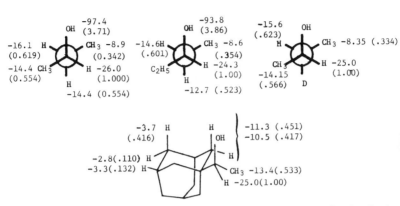

1. Effect of Eu(dpm)$_3$ on pmr chemical shifts of alcohols: Δ_{Eu} values, ratios in parentheses.

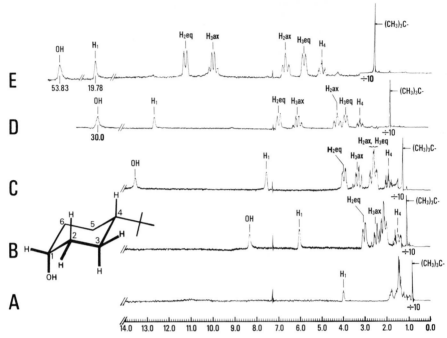

2. 100-MHz pmr spectra of <u>cis</u>-4-<u>tert</u>-butyl-cyclohexanol
    (20 mg, 1.28 x 10⁻⁴ mol) in CDCL₃ (0.4 ml) containing
    various amounts of Eu(dpm)₃: A, 0.0 mg; B, 10.3 mg;
    C, 16.0 mg; D, 33.1 mg; E, 60.2 mg. (Reprinted
    courtesy of Am. Chem. Soc.[7])

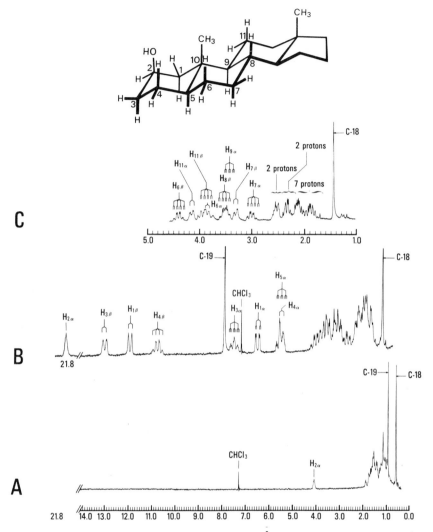

3. Pmr spectra of androstan-2β-ol ($\underset{\sim}{1}$) (20 mg, 0.73 x 10^{-4} mol in 0.4 ml of CDCL$_3$): A, normal spectrum at 100 MHz; B, 100 MHz spectrum of solution of $\underset{\sim}{1}$ containing 40 mg of Eu(dpm)$_3$; C, 220-MHz spectrum of the δ 1.0-5.0 region of spectrum B. (Reprinted courtesy of Am. Chem. Soc.[14])

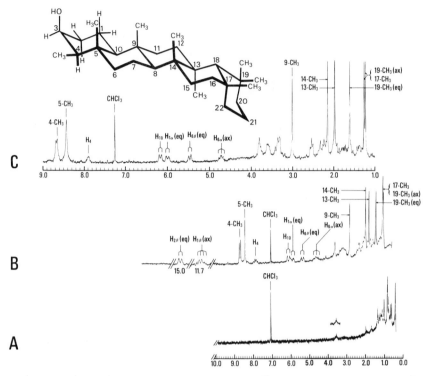

4. Pmr spectra of friedelan-3β-ol ($\underset{\sim}{2}$) (10 mg, 0.24 x 10^{-4} mol in 0.4 ml of $CDCL_3$): A, normal spectrum at 100 MHz; B, 100 MHz spectrum of solution of $\underset{\sim}{2}$ containing 1 mol equivalent of Eu(dpm)$_3$ (17 mg); C, partial 220 MHz spectrum of same solution as in B. (Reprinted courtesy of Am. Chem. Soc.[14])

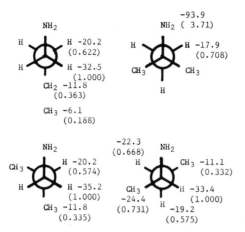

5. Effect of Eu(dpm)₃ on pmr chemical shifts of amines: Δ_{Eu} values, ratios in parentheses.

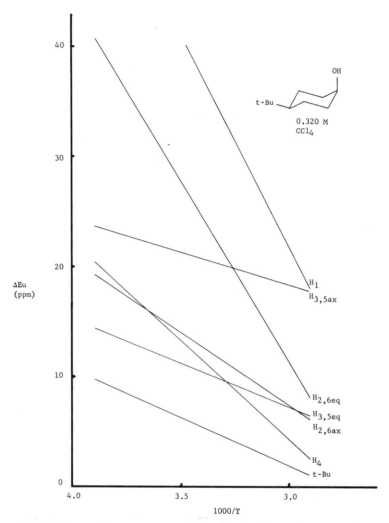

6. Effect of temperature on Δ_{Eu} values for <u>cis</u>-4-<u>t</u>-butylcyclohexanol; plotted as the reciprocal of absolute temperature.

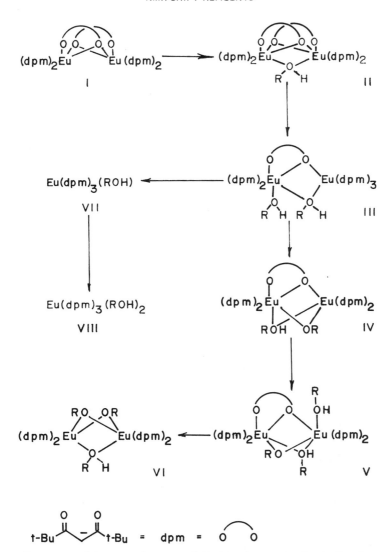

7. Possible complexes of Eu(dpm)₃ and alcohols in
   solution.

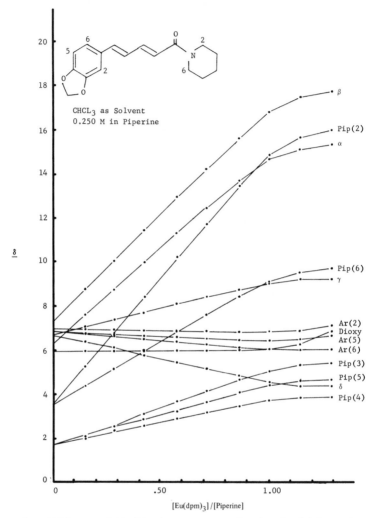

CHCL₃ as Solvent
0.250 M in Piperine

8. Effect of Eu(dpm)₃ on pmr chemical shifts, in chloroform, 0.25 mol in piperine.

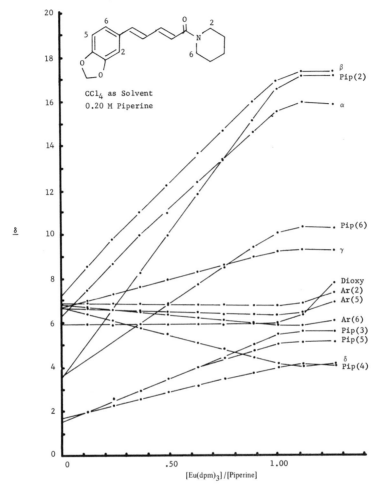

9. Effect of Eu(dpm)₃ on pmr chemical shifts in CCL₄,
    0.20 mol in piperine.

LANTHANIDE-INDUCED ^{13}C NMR SHIFTS (1-2)

G. E. Hawkes, C. Marzin, D. Leibfritz, S. R. Johns
K. Herwig, R. A. Cooper, D. W. Roberts, and
J. D. Roberts

Gates and Crellin Laboratories of Chemistry
California Institute of Technology
Pasadena, California 91109

Our interest in shift reagents is simple — to determine conformations of flexible molecules in solution. And, despite the title of this paper, we will focus on this aspect of the use of lanthanide shift reagents, quite independently of many of the other problems that have arisen in connection with these substances.

Our approach to, and use of, shift reagents, like that of many organic chemists who have similar interests, do not always seem to meet with the full approval of some of the earlier workers in the field. It seems that the general properties of shift reagents would lead one to expect that only the full equations for non-axially symmetric complexes should be able to account for the dipolar shifts that they produce (3). Nonetheless, we have had considerable success with the simple form of the dipolar equation as relates to axial symmetry (3), and to explain how we use it, we need to define some geometrical terms. Figure 1 shows our coordinate system for complexation of a lanthanide chelate with an alcohol (which fits with the rest of our approach by being left-handed). It is convenient to put the complexed atom of the substrate, here, oxygen at the origin, C1 along the +Z axis, C2 in the Y, Z plane. ROM is oxygen-metal distance, $\angle$COM is the carbon-oxygen-metal angle, while θ is a dihedral angle for rotation of the metal around the C1-oxygen bond. Figure 2 shows a magnetic dipole centered on the metal, which makes an angle, χ, with the vector of length R connecting the metal with some

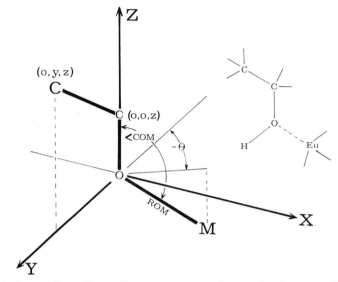

Figure 1.   Coordinate system for calculation of
dipolar interactions.

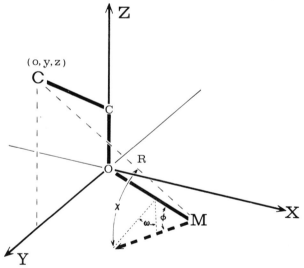

Figure 2.   Location angles for effective magnetic
axis of the lanthanide atom.

atom of interest, in this case, C2. The magnetic effect of the dipole at C2, assuming axial symmetry and freely tumbling molecules, should be proportional to $(3 \cos^2\chi - 1)/R^3$. We define an angle ϕ that the magnetic dipole makes with ROM, and another angle ω, which is a dihedral angle for rotation of the magnetic dipole around the oxygen–metal bond axis. Thus, we have five parameters for the metal: ROM, $\angle$COM and θ which define its position in space, and angles ϕ and ω which determine the position of the magnetic dipole. Other workers with similar approaches have generally considered ϕ to be zero from the beginning (3).

We have written a computer program called CHMSHIFT (4) which varies all five of these parameters in such a way for rigid molecules so as to maximize the correlation coefficient between calculated and experimental shifts. Because, to get conformational information, we need all the experimental points we can get, so we have investigated both proton and ^{13}C lanthanide–induced shifts (5). We have done quite a few molecules, and it seems very significant that $\angle$COM and ROM generally come out with the chemically realistic values of about 130° and 2.7 A, respectively (4,6). This, without arbitrary restraints on any of the five parameters, except the notion that either there is a unique value of θ or, at least, a value which represents a viable average for two or more different conformations (3b, 6b, 7).

Some typical results with the shifts of borneol are shown in Figure 3. Ten proton shifts and eight carbon shifts give a very superior correlation. There are several points of great interest about this correlation. First, the geometrical parameters of the metal are all highly reasonable, with θ coming out so as to put the metal in a favorable steric position. Next, ϕ, is very small, which means that the magnetic dipole is effectively directed along the coordination bond. The angle ω is poorly defined when ϕ is small and will be ignored here. Finding small ϕ provides a great simplification that one should be grateful to Nature for. The final point of interest is that the shifts of C1 and C2 do not correlate well. Either the equations are bad, or some other factor is creeping in. We prefer the latter explanation and, for now, we will note that C1 is too negative and C2 is too positive. Figure 4 shows a corresponding treatment of isoborneol with very comparable results —

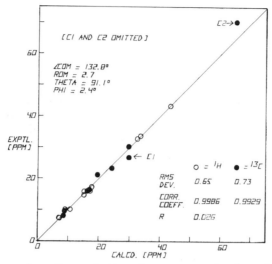

Figure 3. Correlation between calculated and observed proton and carbon shifts for Pr(FOD)₃ and borneol, with C1 and C2 omitted.

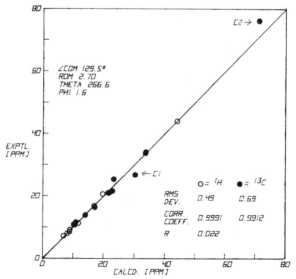

Figure 4. Correlation between calculated and observed proton and carbon shifts for Pr(FOD)₃ and isoborneol, with C1 and C2 omitted.

the geometry is reasonable with θ such as to bring the metal to a sterically favored position, φ is small and, again, Cl is too negative and C2 too positive. These deviations, we believe, arise from contact shifts which turn out to be especially important for amines (6a, 8, 9). Figure 5 shows a CHMSHIFT treatment of *exo*-norbornylamine with Eu(DPM)₃, wherein very large deviations are seen for Cl and C2 (8). The other carbons, except possibly for C3, accord with the proton shifts. Why are Cl (the bridgehead carbon) and C3 so different? Both are β carbons. Our experience is that the more heavily substituted β carbons are especially susceptible to abnormal shifts, and that europium is much worse than praseodymium. Table I demonstrates how, with europium chelates, the 2-carbon of 1-butylamine experiences a wrong-way lanthanide shift, while praseodymium chelates appear to give more or less normal shifts (8). Table II gives similar data for 2-butylamine, the Cl shift, again, being more or less normal, while C3 is wrong-way with europium (8). As before, the praseodymium shifts are in the right direction. Table III shows that the abnormal effect in N-butyl compounds are reduced substantially by substitution at the nitrogen. Tertiary amines seem to give quite normal shifts (8).

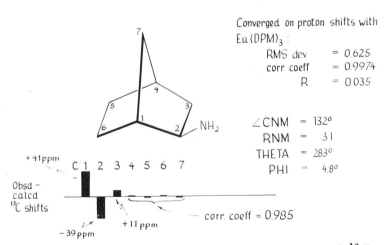

Figure 5. Typical deviation of experimental ¹³C shifts from values calculated from proton shifts on the assumption that dipolar effects predominate.

Table I. Lanthanide–Induced ^{13}C Shifts (in ppm)
for 1–Butylamine.

$$H_2N—CH_2————CH_2————CH_2————CH_3$$

	CH$_2$	CH$_2$	CH$_2$	CH$_3$
Eu(DPM)$_3$	−99. 4	+12. 3	−10. 0	−7. 4
Eu(FOD)$_3$	−94. 5	+25. 4	−5. 3	−5. 3
Pr(DPM)$_3$	+134. 4	+34. 5	+22. 7	+13. 0
Pr(FOD)$_3$	+109. 1	+21. 1	+16. 7	+11. 2
H$^{\oplus}$	+2. 0	+6. 5	0. 0	+0. 1

We interpret the contact–shift contributions at the α
and β carbons and their alternating nature by hyperconju-
gation (Figure 6). Delocalization of a nitrogen electron to
the metal, as shown, would produce a radical electronical-
ly somewhat similar to the allyl radical, with positive spin
density on the end carbons and negative spin density on the
middle carbon. So far, we find no compelling evidence of
contact shifts with hydrogen, so we deem the lower right-
hand resonance form in Figure 6 to be unimportant. The
resonance form at the left should be more important the
more heavily the carbon is substituted, because this would

Table II. Lanthanide–Induced ^{13}C Shifts (in ppm)
for 2–Butylamine.

$$
\begin{array}{c}
\qquad\quad \overset{\displaystyle NH_2}{\overset{|}{}} \\
CH_3————CH————CH_2————CH_3
\end{array}
$$

	CH$_3$	CH	CH$_2$	CH$_3$
Eu(DPM)$_3$	−24. 2	−90. 2	+11. 0	−14. 4
Eu(FOD)$_3$	−12. 7	−81. 9	+23. 6	−8. 9
Pr(DPM)$_3$	+54. 5	+121. 0	+34. 6	+23. 9
Pr(FOD)$_3$	+49. 9	+111. 0	+27. 7	+23. 6
La(DPM)$_3$	+2. 1	~0	−0. 6	+0. 2

Table III.   Europium-Induced ^{13}C Shifts (in ppm)
for Butylamines.

CH₃———CH₂———CH₂———CH₂———				
CH_3———CH_2———CH_2———CH_2———				

Eu(FOD)₃

	CH_3	CH_2	CH_2	CH_2	
	-5.3	-5.3	+25.4	-94.5	—NH₂
Eu(FOD)₃	-4.0	-6.6	+2.7	-107.9	—NHCH₃
	-0.6	-1.7	-2.1	-8.8	—N(CH₃)₂
	-9.3	-10.6	+18.6	-97.0	—NH₂
Eu(DPM)₃	-2.9	-6.9	-22.7	-67.5	—NHBu
	-0.9	-2.4	-11.1	-29.0	—N(Bu)₂

make the form energetically more favorable. Indeed, with
quaternary β carbons, the contact contributions to the ^{13}C
shifts are very large.   Thus, with europium (DPM)₃, C1 of
isobornylamine gives a wrong-way shift of 35 ppm, and C3
of 3,3-dimethyl-2-butylamine is similarly shifted by 45
ppm (9).   Any reasonable correction for the dipolar shift
yields contact contributions of 50 – 80 ppm. Clearly, ^{13}C
shifts need to be interpreted with caution by the dipolar
mechanism.

If one hopes to determine conformations by lanthanide shifts, one had best first try some orienting experiments on simple compounds where one knows about what to
expect (10). We have started with neohexylamine.   This
substance is known from its proton spectrum to be about
three-fourths in the configuration, with the NH₂ group
trans to the *t*-butyl group (11).   One immediate concern is
whether coordination with lanthanide would change the conformational population (3b, 12). Indeed, it does, as judged
by the change in the AA'BB' pattern of the α and β protons
(Figure 7) with europium concentration.   The changes in
proton couplings suggest that the complexed amine essentially has the *trans* configuration at the C1–C2 bond. Knowing this, we next endeavor to locate the complexed lanthanide atom.   The observed ^1H and ^{13}C lanthanide-induced
shifts are shown in Table IV and, again, a large wrong-
way shift is found for C2 with europium. We clearly have

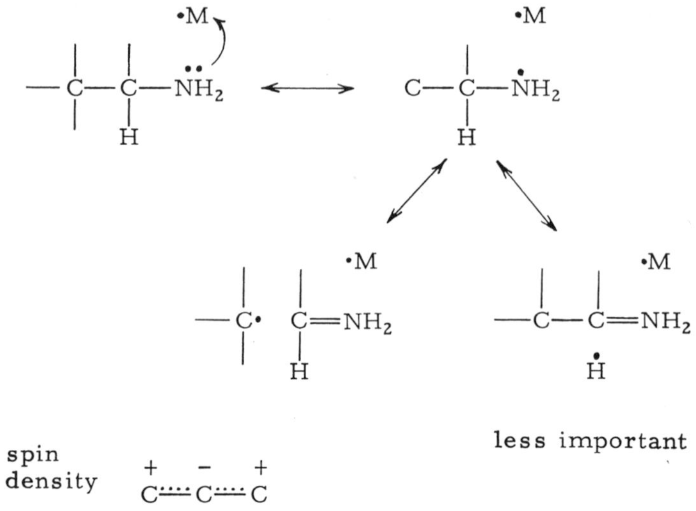

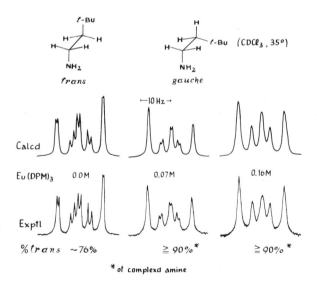

spin
density

$$\overset{+}{C}\cdots\overset{-}{C}\cdots\overset{+}{C}$$

less important

Figure 6. Hyperconjugative resonance forms which account for the difference in sign of the contact contributions at C1 and C2 of amines.

Figure 7. Change in the AA'BB' proton spectrum of neohexylamine with concentration of Eu(DPM)₃.

Table IV. Lanthanide–Induced ^{1}H and ^{13}C Shifts (in ppm) of Neohexylamine with Lanthanide Chelates.

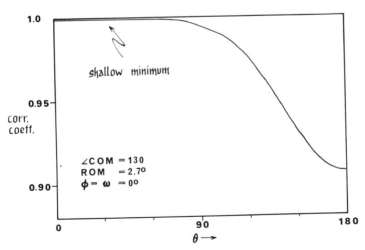

	H1	H2	H4	C1	C2	C3	C4
Eu(FOD)$_3$	−26.4	−18.0	−5.0	−101.1	+24.9	−5.9	−5.9
Pr(FOD)$_3$	+52.6	+32.1	+9.3	+122.9	+20.5	+8.0	+11.9
Yb(FOD)$_3$	−144.4	−66.4	−18.8	−177.7	−73.5	−34.4	−23.2

few parameters to work from, so we assume from our studies of rigid molecules that $\angle$COM $\cong 130°$, ROM $\cong 2.7$ A, and θ and ω both are zero. With these figures, we can attempt to use the proton data to fix θ. Obviously, the ^{13}C data is too likely to be subject to contact interactions to use in this first stage. Figure 8 shows how the correlation coefficient between the calculated and experimental values of the shifts change with θ. The degree of correlation of the proton shifts is quite insensitive to θ from about −80° to +80°. Clearly, these shifts alone are not a good way to fix θ. Similar results have been obtained with praseodymium and ytterbium chelates, and also using cyclopropylamine as substrate. One reason for the difficulty is the predicted

Figure 8. Dipolar correlation of calculated and observed ^{1}H shifts for neohexylamine.

insensitivity of the α-hydrogen shifts to θ. The change of
$(3 \cos^2\chi - 1)/R^3$ as a function of θ for the α hydrogens and
C2 is shown in Figure 9. Both χ and R change drastically
during this rotation, but in such a way as to almost exactly cancel one another. When R is short, χ is large, and
vice versa. This lack of sensitivity is a fairly general
problem for atoms close in. Figure 10 shows how for the
2- and 3-hydrogens of *endo*-norborneol over a range of at
least 120°, the predicted shifts do not vary much with θ.

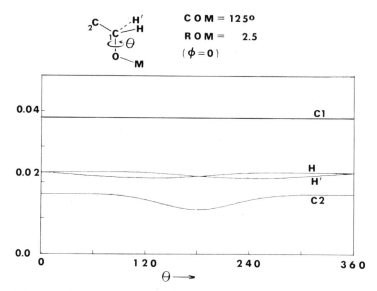

Figure 9. Variation of calculated dipolar effects
at α H's, C1 and C2 for normal lanthanide-substrate
geometry as a function of θ.

Good sensitivity is only possible if θ falls between about
70° and 300°. The ^{13}C shifts are not much help in this
situation. In Figure 11, the calculated contact contributions to the overall ^{13}C shifts are shown as a function of θ.
The contact contributions were computed by using the proton shifts to compute what the dipolar contribution to the
carbon shifts should be — then subtracting these from the
measured values (8). The contact effects are seen to be
rather constant and are also in the directions expected
from our studies of borneols and norbornylamine (4, 8).
The average values shown in Figure 9 are for the range of

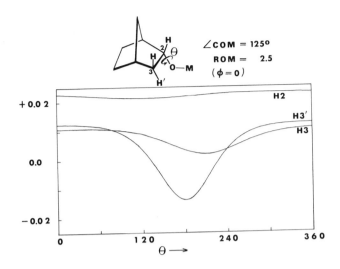

Figure 10. Variation of calculated dipolar effects at H2, H3 and H3' of *endo*-norborneol for normal lanthanide-substrate geometry as a function of θ.

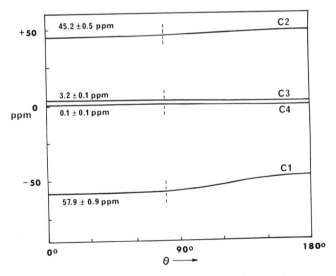

Figure 11. Derived ^{13}C contact shifts for neo-hexylamine with Eu(DPM)$_3$. Average values are for the range of θ from $-80°$ to $+80°$.

139

θ from −80° to +80°. The same procedure applied to europium, praseodymium and ytterbium chelates showed that all three metals display contact shifts which, at C2, fall in the order europium > praseodymium > ytterbium. There is fairly good correlation (Figure 12) between the contact contributions calculated for C2 and the Landé g factor and spin angular momentum of the metals (3b, 13). Similar results, but with smaller contact contributions, are observed for cyclopropylamine. Even with ytterbium, the contact contributions to the C1 and C2 ^{13}C shifts are probably too large to help to fix θ accurately.

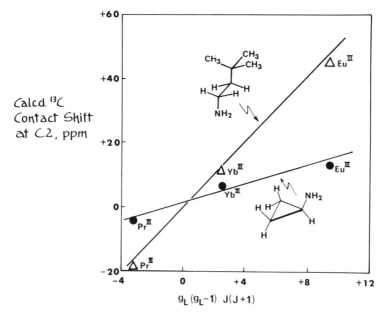

Figure 12. Calculated ^{13}C contact shift contributions at C2 for neohexylamine and cyclopropylamine for different lanthanides.

While this is not an optimistic situation, we should make clear that it is a fairly special one, because we are trying to fix θ with a limited data set. Well-defined values of θ usually come out for rigid molecules like borneol, where there is a large data set and the molecular geometry is such as to spread the atoms out in space. The problem

140

we are trying to solve here is how to find θ with a highly flexible molecule, with just a few close-in points of known geometry, and then proceed to evaluate stepwise the other possible rotational angles in the chain. There are other possible ways to surmount the difficulties in determining θ. R. J. P. Williams has suggested that we try to use the effect of gadolinium on the carbon T_1 relaxation times (3b, 14). The relaxation times should show an R^6 dependence. Figure 13 shows that the R^6 dependence leads to expectation that the ratio of the gadolinium effects on the relaxation times of C1 and C2 should range from about six to unity, as θ changes from 0° to 180°. The experimental ratio of the T_1 effects for 0.025 M Gd(FOD)$_3$ is 7 ± 1.4 at C1 and C2, which agrees best with $\theta = 0°$. This value corresponds to the *trans* configuration of the M-O-C1-C2 bonds, and is the reasonable value of θ on steric grounds. Whether θ can surely be expected to be the same for europium and gadolinium provides another problem, but, at least, the relaxation times can give a check on what other methods might be used to obtain θ.

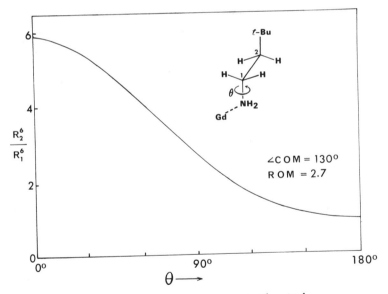

Figure 13. Calculated ratio of R^6_{C1}/R^6_{C2} as a function of θ for gadolinium neohexylamine complex, where R_{C1} and R_{C2} are the Gd-C1 and Gd-C2 distances.

References

1. Contribution No. 4693 from the Gates and Crellin Laboratories of Chemistry.
2. Supported by the Public Health Service, Research Grant No. GM–11072 from the Division of General Medical Sciences, and by the National Science Foundation.
3. (a) W. D. Horrocks, Jr., and J. P. Sipe, III, J. Amer. Chem. Soc., 93, 6800 (1971); (b) J. Reuben, "Progress in Nuclear Magnetic Resonance Spectroscopy", Volume 8, J. W. Emsley, J. Feeney and L. H. Sutcliffe, Eds., Pergamon Press, 1972.
4. G. E. Hawkes, D. L. Leibfritz, D. W. Roberts, and J. D. Roberts, J. Amer. Chem. Soc., 95, 1659 (1973)
5. All of the lanthanide–induced shifts reported here are obtained by least–squares fit for five to six concentrations of lanthanide chelate with correlation coefficients > 0.99 and represent hypothetical shifts at 1 M concentrations.
6. (a) G. E. Hawkes, C. Marzin, S. R. Johns, and J. D. Roberts, J. Amer. Chem. Soc., 95, 1661 (1973); (b) see M. R. Willcott, III and R. E. Davis in this volume, which also has an excellent bibliography of other analyses of lanthanide–shifts in terms of molecular geometries.
7. I. M. Armitage, L. D. Hall, A. G. Marshall, and L. G. Werbelow, J. Amer. Chem. Soc., 95, 1437 (1973).
8. C. Marzin, D. Leibfritz, G. E. Hawkes, and J. D. Roberts, Proc. Nat. Acad. Sci. USA, 70, 562 (1973).
9. Unpublished experiments by G. E. Hawkes and R. A. Cooper.
10. This kind of stricture hardly seems to have been given much consideration in the approach of N. S. Angerman, S. S. Danyluk, and T. A. Victor, J. Amer. Chem. Soc., Soc., 94, 7137 (1972).
11. G. M. Whitesides, J. P. Sevenair, and R. W. Goetz, J. Amer. Chem. Soc., 89, 1135 (1967).
12. See, for example, W. G. Bentrude, H.–W. Tan, and K. C. Yee, J. Amer. Chem. Soc., 94, 3264 (1972).
13. W. B. Lewis, J. A. Jackson, J. F. Lemons, and H. Taube, J. Chem. Phys., 36, 694 (1962).
14. Cf., H. Sternlicht, J. Chem. Phys., 42, 2250 (1965).

ASSESSMENT OF THE PSEUDOCONTACT MODEL
VIA AGREEMENT FACTORS

Raymond E. Davis
Department of Chemistry, University of Texas at Austin
Austin, Texas   78712

M. Robert Willcott, III
Department of Chemistry, University of Houston
Houston, Texas   77004

Lanthanide shift reagents (LSR) owe their popularity
to the fact that they alter high resolution nuclear magnetic resonance spectra in an interpretable fashion.   The
qualitative aspects of the LSR experiment can be sufficient
to remove structural ambiguities, but several research
groups have developed quantitative techniques for the analysis of the data (1).   The published quantitative procedures differ primarily in the details of the construction
of a computer program.   Presumably all of the procedures
will lead to similar conclusions about the topology of the
substrate since each of the procedures seems to succeed
when applied to molecules of interest to that particular
research group.   We have found, in the course of developing
and assessing quantitative approaches to interpreting the
LSR experiment in terms of structure, that a critical comparison of the various schemes for simulation of LSR spectra leads us to review several major aspects of the problem.
In this paper we illustrate some simulated LSR spectra of
simple molecules, evaluate the appropriateness of the dipolar axially symmetric equation, and assess the usefulness
of this quantitative approach as a tool for structure
studies.

It is important to be clear about what is meant by
"usefulness" of the approach as a structural tool.   We
should first note that we do not <u>determine</u> structures from
LSR data, in the sense of deriving positional or structural
coordinates from the data; rather, we <u>propose</u> the various

structural possibilities, and then use the LSR data to test these proposals, and hopefully to validate or at least reject them. Thus, "usefulness" may not have to mean rigorous mathematical applicability.

In our view, "usefulness" as a structural tool means: (i) the structural conclusions will not be critically dependent on details of data collection, data reduction, etc. --LSR data should be usable as 1:1 mole ratios, relative slopes, etc.; (ii) the method will work for different nuclei --^{1}H, ^{13}C, ^{19}F, etc.; (iii) the method will work for many (all?) substrates; (iv) the method is relatively insensitive to assumed lanthanide position, since we rarely would have accurate *a priori* knowledge of this position; and (v) the method is relatively insensitive to small errors in the description of the substrate model. Certainly a sensitivity in this description to only about 0.2-0.3 A would be sufficient to distinguish among most structural or geometrical isomers, while a much greater sensitivity would require not easily attainable accuracy in the model description. Further, it must always be borne in mind that in most cases we are faced with a paucity of data, 3 to 15 observations being typical.

Any quantitative approach to fitting LSR spectra has three aspects: (i) the physical-mathematical model (On what description of the phenomena being observed will we base our attempts to fit the observations?); (ii) the operational or computational model (What will be our method for carrying out the attempts at fitting?); and (iii) the assessing function (How can we tell how well our attempts to fit the observed data have succeeded?).

Most attempts (1,2) at fitting the LSR data use, as the physical-mathematical model, a dipolar field, either axially symmetric or effectively axially symmetric (Figure 1). For such a situation, with the principal magnetic axis of the lanthanide taken to be collinear with the lanthanide-base bond, the dipolar (pseudocontact) contribution to the lanthanide induced shift (LIS) of atom i is given by the McConnell-Robertson (3) relationship:

$$\text{Pseudocontact contribution} = k \left[\frac{3 \cos^2 \theta_i - 1}{r_i{}^3} \right] \quad \text{(eq. 1)}$$

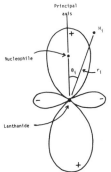

Figure 1. The dipolar, axially symmetric field.

The process of fitting the data is depicted in Figure 2.

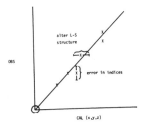

Figure 2. Fitting of calculated to observed LIS values.

We place the model of the substrate in the field of Figure 1, with the base site on the principal axis. We then vary the orientation of the substrate model until the calculated LIS values, plotted against the observed ones, give as good a straight line as possible. Then, since the observed LIS data are on some unknown scale, relative to equation 1, the constant scale factor in equation 1 must be determined. It may also be necessary to not constrain the line of Figure 2 to pass through the origin, in order to allow for a constant additive error in each observation --e.g., lack of an internal standard, resulting in bulk susceptibility changes. This fitting procedure requires at least 4 parameters (three orientational, at least one scale).

Several alterations to this simplest model are possible, e.g.: (i) the principal magnetic axis need not be taken along the lanthanide-base bond (1j); (ii) the axial symmetry restriction may be released (4); (iii) several non-symmetry-related lanthanide sites may be taken into account, with a weighted average effect being calculated (e.g., 1c, 1m); (iv) Fermi contact shift contributions may be accounted for by factoring of the observed LIS data. However, since any of these alterations requires additional adjustable parameters to be derived from the small data set available, it seems mandatory to use the most economical reasonable model which still is useful for structure assessment. We have chosen to use the simple model, previously described, of a dipolar, effectively axially symmetric field with the principal magnetic axis along the lanthanide-base bond.

Our computational approach (1f) has been to hold the substrate model coordinates fixed, and to then systemati- cally vary the lanthanide position over all reasonable space near the base site, by moving it incrementally over the sur- faces of closely spaced spheres. The location of the lan- thanide on each sphere is described in terms of the two angles ρ, the colatitude, and ϕ, the aximuth. At each posi- tion, the variable term in equation 1 is calculated for each atom. This set of calculated values is then scaled by least squares against the observed LIS values, to yield a set of calculated shifts.

Several possible functions have been proposed and used for assessing agreement between observed and calculated values, including standard deviation and correlation coef- ficients. We have found it convenient to use the quantity

$$
R = \left[\frac{\Sigma w_i \ (obs_i - cal_i)^2}{\Sigma w_i \ (obs_i)} \right]^{1/2} \qquad \text{(eq. 2)}
$$

where w_i is a weight applied to the observation i. The quantity R is essentially a normalized standard deviation, and has the attractive feature that similar values of R have about the same meaning for different sets of data (of about the same size), even if the data sets are scaled quite dif- ferently. It is to be noted that a minimum R value for a model corresponds to a least-squares minimum for that model, whether the set of parameters giving that minimum was de- rived by conventional least squares techniques, by inter- ative convergence methods, or by brute force systematic variation of parameters, as in our computational scheme.

This minimum R value for various models may also be used in a statistical testing scheme, as described elsewhere (1g,5).

The fit of calculated to observed LIS values may be exemplified by the set of three isomeric acrylonitriles (Table 1). Observed values are relative LIS indices obtained with $Yb(DPM)_3$; calculated values were all obtained with the

Table 1. Relative LIS values, OBS/CAL, for the set of three isomeric methylacrylonitriles; $Yb(DPM)_3$, ^1H.

	1-methyl	cis-2-methyl	trans-2-methyl
CH_3	8.17/8.21	6.59/6.78	4.02/3.96
H_1	---	10.00/10.04	10.00/9.89
H_2-cis	10.00/10.22	---	8.20/8.36
H2-trans	6.94/6.54	5.83/5.51	---
R (%)	3.1	2.8	2.6

lanthanide in the range 2.6-2.9 A and within 5° of the nitrile axis. Within this range, alterations in the lanthanide position lead to small changes in agreement, well within the errors in observations. When attempts are made to fit the incorrect isomer to any set of observed LIS values, much poorer agreement is obtained (6). Since only three observations are available, it is possible to fit any of the three models to any of the three sets of LIS data, but at the expense of a totally unreasonable lanthanide position. This raises the general question of the existence of homometric sets--i.e., 2 or more structures which could give the same set of LIS observations. While such cases can be designed on paper, they are probably rare in practice, and could usually be distinguished by use of information other than the LSR experiment. For example, Paasavirta has reported the LIS data for a set of isomeric [2.2.1] heptanols, structurally quite similar, in which the LIS data differ sufficiently to permit one to distinguish among the various isomers (7).

Another possible difficulty is shown by the data for pyridine with several different lanthanides in Table 2, in which observed LIS data are scaled to H_γ = 1.0 for comparison. The calculated values for lanthanide 2.7 A from N

Table 2.   Pyridine-LAN(DPM)$_3$ relative LIS values; ^{1}H.

	Eu (obs)	Pr (obs)	Yb (obs)	Yb (cal)*
H_α	3.16	3.52	3.22	3.23
H_β	1.10	1.25	1.26	1.21
H_γ	1.00	1.00	1.00	1.00

*Calculated for N-Yb = 2.7 A, angle C-N-Yb = 120°;
R = 1.3%

along the symmetry axis of the ring   show excellent agreement with the Yb(DPM)$_3$ data (R = 1.3%). Inspection of the Pr(DPM)$_3$ data reveals a discrepancy in the H_α/H_β ratio, while the H_β/H_γ ratio is out of line for the Eu(DPM)$_3$ complex. These different ratios could be fit by moving the lanthanide to other, less reasonable positions, but it seems more reasonable to assume that the lanthanide atom occupies the same position in all three complexes. Another, more likely explanation, is the incursion of a sizable degree of Fermi contact shift into the latter two spectra. This possibility has been investigated by a number of workers (8).

A convenient opportunity to study the effects of contact shift in several nuclei of the same molecule is afforded by the molecule 3-fluoropyridine (9). The substitution of fluorine for hydrogen at position 3 causes little steric alteration, while allowing a much greater sensitivity to contact shift (F>C>H). Since a good fit is possible for Yb(DPM)$_3$ with pyridine, let us first consider the observed and calculated values for 3-fluoropyridine-Yb(DPM)$_3$, shown in Figure 3 (calculated values for Yb on "symmetry" axis, N-Yb distance = 2.6 A, R = 3.2% including all nuclei). This very good fit with the simple dipolar model suggests that Yb(DPM)$_3$ induces little contact shift in the LIS values for any of the three kinds of nuclei, and that either the ^{1}H LIS spectrum, the ^{13}C LIS spectrum, or the combined ^{1}H and ^{13}C data would have been suitable for structural interpretation.

A marked contrast is seen in the case of 3-fluoropyri-

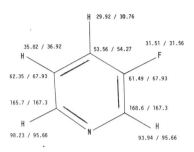

Figure 3. LIS data for 3-fluoropyridine, Yb(DPM)$_3$. Ratios are OBS/CAL, with CAL calculated for Yb-N = 2.6 A, angle C-N-YB = 120°. R = 3.2%.

dine-Eu(DPM)$_3$ (Figure 4). The calculated values have been obtained for the same position as for Yb(DPM)$_3$, and then scaled to fit the observed value for H$_4$, the nucleus far-

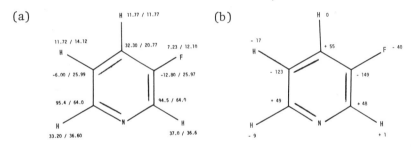

Figure 4. LIS data for 3-fluoropyridine, Eu(DPM)$_3$. (a) OBS/CAL, with CAL calculated for N-Eu = 2.6 A, angle C-N-Eu = 120°, scaled to H$_4$. (b) Contact/pseudocontact ratio, as percentages.

thest removed from the base site, expected to be least affected by contact shift. We particularly note the poor fit at C$_3$ and C$_5$, where the observed shift is the opposite direction from that calculated with the dipolar model. It is seen that the agreement for ^{1}H nuclei is still rather good, while the observed and calculated values are in marked disagreement for both ^{13}C and [19]F. It is convenient to generate a difference spectrum, i.e., the difference between observed LIS value and calculated LIS value based on the dipolar model, as a measure of contact shift contribution. This contact shift, expressed as a percentage of calculated dipolar shift, is shown in figure 4b. We see that the "average" contact shift in the ^{13}C LIS spectrum due to Eu(DPM)$_3$ is about 85% of the dipolar, or pseudocontact contribution. Thus, the ^{13}C spectrum would have been quite unsuitable for any possible structural study, while the ^1H spectrum remains free enough of contact shift to have allowed a fair agreement with the dipolar model.

We have investigated the trends in the relative contact/pseudocontact contributions for various lanthanides in the series iso-quinoline-LAN(DPM)$_3$, where LAN = Eu, Nd, Er, Tb, Ho, Pr, Dy and Yb (10). In each case, the lanthanide position and scale factor derived from the ^1H LIS data were used to calculate the pseudocontact contribution at each carbon atom. This calculated value was then subtracted from the observed ^{13}C LIS value, and the result, taken as the contact shift, was expressed as a percentage of the pseudocontact shift. The results, averaged for the five carbon atoms in the nitrogen ring, are shown in Table 3. The trend in this table is in agreement with the predicted pseudocon-

Table 3. Agreement factors, average values for ^{13}C contact contributions, isoquinoline vs. LAN(DPM)$_3$.

| | LAN | | | | | | | |
	Eu	Nd	Er	Tb	Ho	Pr	Dy	Yb
R_H	3.0	4.8	1.1	1.5	1.3	2.0	1.8	1.4
$R_{C,H}$	47.7	46.8	17.6	26.9	33.3	11.6	30.0	4.0
Contact*	80	70	25	21	19	15	13	5

*Contact shift, expressed as a percentage of pseudocontact shift, averaged for the 5 carbon atoms of the nitrogen ring.

tact/contact shift contributions calculated by Reuben (11). This table also shows the best obtainable R value for the ^1H LIS spectrum alone, and that for the combined ^1H and ^{13}C LIS spectra. By either measure, the spectra obtained with Eu(DPM)$_3$ give poorest agreement, and those with Yb(DPM)$_3$ give best agreement with the dipolar model. This strongly suggests that Yb(DPM)$_3$ is a more suitable shift reagent than is Eu(DPM)$_3$ where quantitative interpretation of the data is to be carried out, and that the dipolar calculations alone would be quite inadequate for Eu(DPM)$_3$ ^{13}C LIS spectra. It further suggests that, while collection of matched proton and carbon spectra with Yb(DPM)$_3$ could help overcome statistical problems due to scarcity of data, use of such matched Eu(DPM)$_3$ spectra with the dipolar structural interpretation would be greatly in error.

Sensitivity to contact shift contributions, as well as to lanthanide position determination, has been investigated (12) with the matched ^{1}H and ^{13}C LIS spectra for *endo*-norborn-5-en-2-ol with Yb(DPM)$_3$ and with Eu(DPM)$_3$ as the LSR. Agreement between observed and calculated ^{1}H LIS spectra for Yb(DPM)$_3$ is excellent, as shown in Figure 5. Dependence of

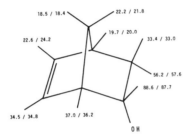

Figure 5. LIS data for *endo*-norborn-5-en-2-ol, Yb(DPM)$_3$, H. Ratios are OBS/CAL. R = 2.0%, O-Yb = 2.7 A.

R on lanthanide position is conveniently displayed using contour lines on a map projection (1f) such as those shown in Figure 6. These plots show contours of constant R as a function of ρ and ϕ with each plot representing a shell of constant assumed lanthanide-base distance. Such representations are useful for revealing the presence of multiple or unusually sharp minima, both of which can be indications of some problem in the calculation (e.g., incorrect assignments).

Comparison of the three plots of Figure 6 shows the relative insensitivity (to 0.2-0.3 A) of the calculation to assumed lanthanide position, either in terms of the lanthanide base distance or of ρ or ϕ. For ease of comparison among the remaining plots to be presented, each plot is marked with a cross at the position of best agreement for Yb(DPM)$_3$, ^{1}H, 2.7 A. Figure 7, the agreement factor plot for Eu(DPM)$_3$, ^{1}H, 2.7 A, is quite similar to the best fit plot for Yb(DPM)$_3$.

(a)

(b)

(c)

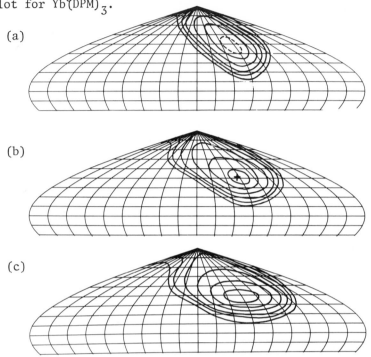

Figure 6. Agreement factor plots for *endo*-norbornenol, Yb(DPM)$_3$ ^{1}H. Outer contour is at 18%, increments of 3%. Inner contour is 3% where solid, 4% where dashed. (a) O-Yb = 2.4 A. (b) O-Yb = 2.7 A. (c) O-Yb = 3.0 A. The cross in this and subsequent plots marks the position of best agreement for Yb(DPM)$_3$ ^{1}H values.

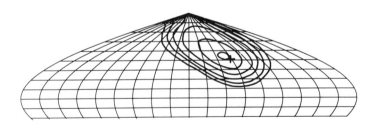

Figure 7. Agreement factor plot for *endo*-norbornenol, Eu(DPM)$_3$, ^{1}H. Contours as in Figure 6.

The dependence of R factor sensitivity on quality of the data is investigated by generating an "ideal" set of ^{1}H data, using the calculated values from the Yb(DPM)$_3$ 2.7 A calculation as "observed" values. The agreement factor plot, Figure 8, is virtually identical with that for the real data, Figure 6b, except, of course, that the minimum R

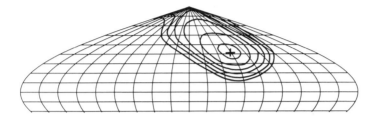

Figure 8. Agreement factor plot for *endo*-norbornenol, using ideal data as described in text. Contours as in Figure 6.

value obtained corresponds to perfect fit. The close agreement of these two plots clearly indicates the limiting returns to be gained, past a point, from continued polishing in data reduction.

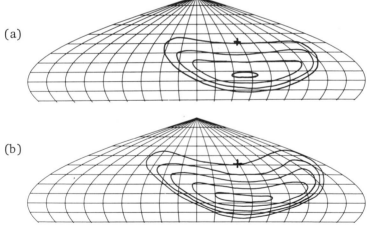

(a)

(b)

Figure 9. Agreement factor plots for *endo*-norbornenol, ^{13}C, O-LAN = 2.7 A. (a) Eu(DPM)$_3$. Outer contour at 18%, inner contour at 9%. (b) Yb(DPM)$_3$. Outer contour at 18%, inner contour at 6%.

Agreement factor plots for ^{13}C LIS spectra are shown in Figure 9. In both these plots, the region of minimum R is much more diffuse than for ^1H, is badly displaced to a position of severe steric crowding, and does not reach as good an agreement. Presumably this poor fit is again due to the presence of some contact shift contribution to the ^{13}C LIS spectra. Since contact shift would be expected to be worst for atoms closest to the base site, these calculations were repeated omitting the LIS value for C_2. Figure 10a shows that, while the minimum R for $Eu(DPM)_3$ now is found at about the correct lanthanide position, much poorer agreement is obtained (minimum R = 14.2%), suggesting that contact shift contributions are still important for

(a)

(b)

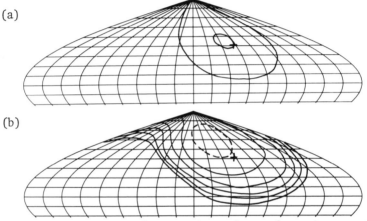

Figure 10. Agreement factor plots for *endo*-norbornenol, ^{13}C, omitting C_2, O-LAN = 2.7 A. (a) $Eu(DPM)_3$, outer contour at 18%, inner contour at 15%. (b) $Yb(DPM)_3$, outer contour at 18%, dashed contour at 4%.

other carbon atoms in the molecule. For $Yb(DPM)_3$, ^{13}C, 2.7 A (Figure 10b), the region is still rather diffuse but is more correctly placed, and a minimum R value of 3.9% is obtained, compared to the 2.0% for ^1H. Clearly, most of the contact shift difficulties with the carbon spectrum have now been removed by omission of C_2.

In order to demonstrate the sensitivity of the fit to incorrect stereochemical models, we have also attempted to fit the *endo*-norbornenol $Yb(DPM)_3$ ^1H, $Eu(DPM)_3$ ^1H, and

Yb(DPM)$_3$ ^{13}C (omitting C$_2$) data with the model coordinates for *exo*-norbornenol. In none of these cases could agreement better than 19% be obtained, with even these minima frequently corresponding to impossible lanthanide positions. This is consistent with our experience for a wide variety of rigid or mostly rigid compounds (including alcohols, ketones, aldehydes, epoxides, ethers and nitrogen aromatics), that incorrect isomers usually do not give agreement better than R≈ 12-15%, while correct models exhibit a smooth minimum in agreement factor of about 6% or better.

In summary, while the dipolar model with effective axial symmetry is surely not a complete description of the lanthanide-substrate complex, it does give excellent descriptions of ^{1}H LIS spectra, apparently for the entire range of lanthanide shift reagents studied (13). The agreement obtained with this model is insensitive to data reduction methods, relatively insensitive (0.2-0.3 A) to assumed lanthanide position or to error in model coordinates, but sensitive to signal assignments and to substrate stereochemistry. It can thus be used with confidence for ^{1}H LIS studies of substrate structure. It is equally clear that the presence of large amount of contact shift contribution in most ^{13}C and ^{19}F LIS spectra (with Yb(DPM)$_3$ a possible exception) limits the applicability of the dipolar model for these spectra.

Copies of the computer programs used in these calculations are available (14).

ACKNOWLEDGEMENTS: This work was generously supported by the Robert A. Welch Foundation, through Grants F-233 and E-183. MRW expresses his thanks to the John Simon Guggenheim Memorial Foundation for a fellowship, and to the California Institute of Technology Division of Chemistry for providing a stimulating environement in which to pursue these studies.

155

References

1.a) J. W. ApSimon and H. Beierbeck, Tetrahedron Letts.,
581 (1973).
  b) N. S. Angerman, S. S. Danyluk and T. A. Victor, J.
Amer. Chem. Soc., 94, 7137 (1972).
  c) I. M. Armitage, L. D. Hall, A. G. Marshall and L. G.
Werbelow, J. Amer. Chem. Soc., 95, 1437 (1973).
  d) J. Briggs, F. A. Hart, G. P. Moss and E. W. Randall,
Chem. Commun., 364 (1971).
  e) J. Briggs, F. A. Hart and G. P. Moss, Chem. Commun.,
1506 (1970).
  f) M. R. Willcott, R. E. Lenkinski and R. E. Davis,
J. Amer. Chem. Soc., 94, 1742 (1972).
  g) R. E. Davis and M. R. Willcott, J. Amer. Chem. Soc.,
94, 1744 (1972).
  h) P. V. Demarco, B. J. Cerimele, R. W. Crane and A. L.
Thakkar, Tetrahedron Letts., 3539 (1972).
  i) J. Goodisman and R. W. Mathews, Chem Commun., 127
(1972).
  j) G. E. Hawkes, D. Leibfritz, D. W. Roberts and J. D.
Roberts, J. Amer. Chem. Soc., 95, 1659 (1973).
  k) T. Heigl and G. K. Mucklow, Tetrahedron Letts., 649
(1973).
  1) P. H. Mazzocchi, H. L. Ammon and C. W. Jameson,
Tetrahedron Letts., 573 (1973).
  m) J. J. Uebel and R. M. Wing, J. Amer. Chem. Soc.,
94, 8910 (1972).
  n) S. Farid, A. Ateya and M. Maggio, Chem. Commun., 1285
(1971).
2.a) H. Huber, Tetrahedron Letts., 3559 (1972).
  b) J. M. Briggs, G. P. Moss, E. W. Randall and K. D.
Sales, Chem. Commun., 1180 (1972).
3.   H. M. McConnell and R. E. Robertson, J. Chem. Phys.,
29, 1361 (1958).
4.a) W. DeW. Horrocks and J. P. Sipe, Science, 177, 994
(1972).
  b) C. L. Honeybourne, Tetrahedron Letts., 1095 (1972).
  c) B. Bleaney, C. M. Dobson, B. A. Levine, R. B. Martin,
R. J. P. Williams and A. V. Xavier, Chem. Commun.,
791 (1972).
5.   W. C. Hamilton, "Statistics in Physical Science",
Ronald Press, New York, N. Y., 1964, pp. 157-162.

6. R. E. Davis, M. R. Willcott, R. E. Lenkinski, W. von E. Doering and L. Birladeanu, submitted for publication.

7. J. Paasavirta and P. J. Malkonen, Suom Kemist., 44, 230 (1971).

8.a) W. DeW. Horrocks and J. P. Sipe, J. Amer. Chem. Soc., 93, 6800 (1971).

   b) M. Kainosho, K. Ajisaka and K. Tori, Chemistry Letters, 1061 (1972).

   c) G. E. Hawkes, C. Marzin, S. R. Johns and J. D. Roberts, J. Amer. Chem. Soc., 95, 1661 (1973).

   d) M. Hirayama, E. Edagawa and Y. Hanyu, Chem. Commun., 1343 (1972).

   e) B. F. G. Johnson, J. Lewis, P. McArdle and J. R. Norton, Chem. Commun., 535 (1972).

   f) G. H. Wahl and M. R. Peterson, Chem. Commun., 1167 (1970).

   g) R. J. Cushley, D. R. Anderson and S. R. Lipsky, Chem. Commun., 636 (1972).

   h) P. Kristansen and T. Ledaal, Tetrahedron Letts., 4457 (1971).

   i) Z. W. Wolkowski, C. Beaute and R. Jantzen, Chem. Commun., 619 (1972).

9. J. D. Roberts, G. E. Hawkes, M. R. Willcott and R. E. Davis, unpublished results.

10. O. A. Gansow, P. A. Loeffler, R. E. Davis, M. R. Willcott and R. E. Lenkinski, J. Amer. Chem. Soc., in press.

11. J. Reuben, paper presented during this symposium.

12. O. A. Gansow, P. A. Loeffler, R. E. Lenkinski, M. R. Willcott and R. E. Davis, unpublished results.

13. A few cases have been reported of obviously large contact shift contributions in ^{1}H spectra, especially with Eu(FOD)$_3$. See, e.g., reference 8e.

14. Inquiries should be directed to R. E. Davis.

CONFIGURATIONAL ASSESSMENT OF CONFORMATIONALLY MOBILE MOLECULES BY THE LIS EXPERIMENT

M. Robert Willcott, III
Department of Chemistry, University of Houston
Houston, Texas   77004

Raymond E. Davis
Department of Chemistry, University of Texas at Austin
Austin, Texas   78712

Static lanthanide-substrate structures have been used for the interpretation of lanthanide-induced chemical shifts (LIS) both in the literature (1) and during the course of this symposium (2). We have suggested that approximations in the mathematical model and errors in the measurement of the induced shift limit the accuracy of the lanthanide (L-S) descriptions to about .3 A. One explanation for this limited accuracy is that the experimental shift perturbations used to deduce the structure of the rigid L-S array result from several dynamic processes, all of which are fast on the nmr time scale. Typical dynamic events are the movement of the lanthanide relative to the substrate, conformational changes in the substrate, and combinations of these motions. A computational method which explicitly considers the movement of the lanthanide relative to the substrate has been published by Armitage, Hall, Marshall, and Werbelow (4). Detailed discussions of methods for the determination of the best conformation for non-rigid substrates relating to a fixed lanthanide position have been provided by Danyluk (5), Williams (6) and their colleagues. Other more fragmentary communications have outlined methods for determining conformations in simple molecules (7). It is clear to us that any attempt to include all possible dynamic events for any single lanthanide-substrate investigation will lead to a variety of solutions since the mathematical form of the analysis will become underdetermined. In the worst cases, the interpretation of the LIS behavior of a substrate becomes ambiguous (8). In this paper we will summarize some of our recent experience in the use of lanthanide–induced shifts to determine conformational preference in flexible molecules.

Arends and Helboe of the Royal Danish School of Pharmacy, Copenhagen, have provided us an interesting set of experimental data for xanthone, the four different methoxyl isomers, and 1-xanthol (9). In every case, the experimental shifts were obtained by sequential addition of Eu(DPM)$_3$ to CDCl$_3$ solutions of the substrate, and data reduction to a 1:1 molar equivalent value was accomplished by a least–squares fit of the L/S ratio *vs.* frequency. Internal coordinates for the xanthone substrates were adapted from the X–ray structures of related compounds (10). Routine employment of the PDIGM program (3) permitted us to replicate the LIS spectrum of the parent compound with an agreement factor of less than 1% (Figure 1).

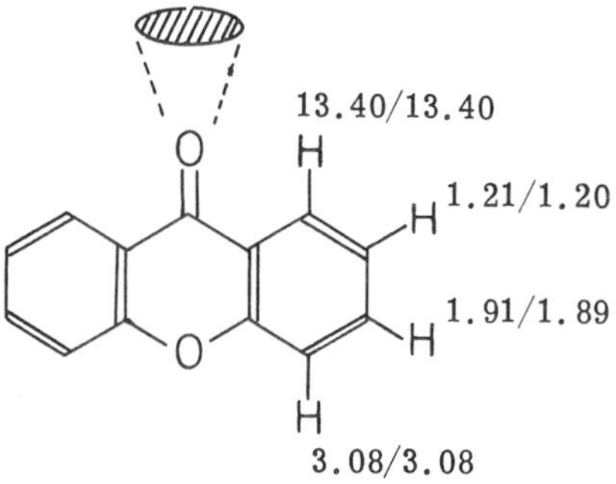

Figure 1. Best fits of observed/calculated LIS spectra for xanthone.

The cross–hatched area in the figure is a graphical representation of the region where agreement factors of less than 3% were found. The region extends from 2.4 to 2.6 A and *ca.* 15° either side of the carbonyl axis. We note that the imprecision in fixing the L–S structure, *i.e.*, the resolution of the experiment, is about .3A. Even though the virtual linearity of the C--O--Eu array was surprising at first sight, we subsequently realized that very little is known about the geometry of complexation of carbonyl groups in the LIS experiment. The fact that the region of best fit is also the region of minimum steric interaction is

160

consistent with the C--O--Eu collinear description for the xanthones. Mazzochi and Ammon have recently discussed complexation of lanthanide shift reagents at carbonyl groups and reached a similar conclusion (11) about collinearity from Wolkowski's fluorenone-Yb(DPM)₃ data (12). Thus, the xanthone-Yb(DPM)₃ structure in Figure 1 can be employed to explore the conformations of the methoxyl groups in the substituted compounds.

Testing all possible L–S structures for 4–methoxyxanthone, using the LIS data for the 7 ring protons, localized the europium in the cross–hatched region in Figure 2(a). If the region contains all europium positions for

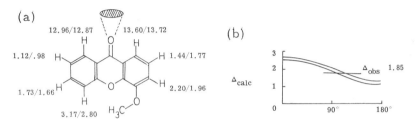

Figure 2. (a) Best fit of observed/calculated LIS spectra for 4–methoxyxanthone; (b) calculated methoxyl shift *vs*. O–methyl position.

agreement factors of less than 5%, it has the same geometric description as that for xanthone. Thus, introduction of the 4–methoxyl substituent caused only a minor perturbation of the L–S structure. Sets of internal coordinates for the methoxyl group were generated as a function of rotation about the ring–oxygen bond. We chose 0° rotation so that the methoxyl framework was coplanar with the aromatic ring, and close to the ether linkage of the xanthone. The LIS shift for the methoxyl group was found by fixing the europium position and computing the methyl shift, scaled to the 7 ring resonances, as a function of the angle of rotation about the ring–oxygen bond. The result is shown as a graph in Figure 2(b). The two curves represent the extreme values of Δ_{calc} as the europium location was varied throughout the cross–hatched region.

Several points merit comment. The distance of the methyl group from the complexation site is great enough that Δ_{calc} is insensitive to the location of the Eu atom and all of the calculated values span the experimental observation. We can include the external evidence that anisoles should have at least two-fold rotational barriers and interpret this LIS result in terms of an approximate 1:1 mixture of 0 and 180° rotamers. This stretches the method to the limit since we have described two populations based on but one experimental observation. More than this we dare not do.

The same treatment was applied to 3-methoxyxanthone, with the 0° rotamer being defined as shown in Figure 3 (a). We found that when the europium was located in the

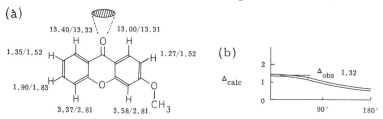

Figure 3. (a) Best fit of observed/calculated LIS spectra for 3-methoxyxanthone; (b) calculated methoxyl shift *vs*. O-methyl position.

same region as for xanthone, the agreement factor was less than 5%. Computation of the methyl shift was straightforward, and is shown in Figure 3 (b). In this case, the observed LIS value of the methyl group coincided with the highest computed values. One interpretation of this result is that only the 0° rotamer is appreciably populated. This behavior is strange enough that we suspect the correct explanation must reside elsewhere. Even so, we have reached the limit for the interpretation of the LIS data for 3-methoxyxanthone. If a more elaborate description of the conformation of the methoxyl group is desired, then more elaborate experiments will have to be conducted.

Treatment of 2-methoxyxanthone in the same manner produces the summaries in Figure 4 (a) and (b). The 0° rotamer is the one illustrated on the formula. The calculated methyl shift in this arrangement is downfield, as it is for the other nuclei in the molecule. When the methyl group is rotated 60 to 80°, an upfield shift results. The

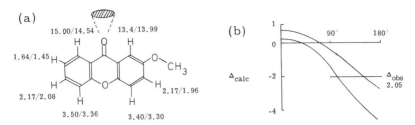

Figure 4. (a) Best fit of observed/calculated LIS spectra for 2-methoxyxanthone; (b) calculated methoxyl shift *vs.* O-methyl position.

proximity of the methyl group to the carbonyl group causes a much greater sensitivity of Δ_{calc} to both rotation of the methyl and location of the europium. As before, the two curves represent the extreme values calculated for all reasonable europium locations. The observed shift is opposite in sign to the other shifts in the molecule and falls in the center of the range of Δ_{calc}. We can suggest that both the 0 and 180° rotamers are about equally populated, but a more precise evaluation is not possible at this time.

Systematic understanding of the xanthones was interrupted by the 1-methoxyl compound. The experimental observations, shown in Figure 5, could not be reproduced

Figure 5. Observed LIS values for 1-methoxyxanthone.

when europium was coordinated to the carbonyl group; the minimum agreement factor was 15%. The situation improved when the oxygen of methoxyl was used as the coordination site, but the minimum agreement factor was still 10%. We suspected the orientation of the principal

163

magnetic axis (13) might be causing these poor fits, so we arbitrarily moved the xanthone in our cartesian coordinate system until a good agreement (less than 4%) was found. This occurred when the effective principal magnetic axis passed through a point equidistant between the two oxygen atoms. This L-S geometry, if correct, suggests that the substrate is behaving as a bidentate ligand. The geometric arrangement of the oxygen atoms is analogous to the arrangement in the ligands used to prepare shift reagents. The importance of the methyl group in the compound was clearly pointed out by Arends' observation that attempts to take the LIS spectrum of 1-xanthol in the presence of Eu(DPM)$_3$ led to an immediate precipitate, and the supernatant solvent contained dipivaloylmethane. The interpretation that the bidentate ligand causes the difficulty in simulating the LIS spectrum of the 1-methoxyl isomer is appealing. We are looking for other cases to see if the problem is general.

We now turn our attention to a somewhat different problem — that of understanding the topology of cyclobutanes via the LIS experiment. We have joined Doering and Sachdev in studying *cis*- and *trans*-1-cyano-2-vinylcyclobutane. Formulas for these molecules along with the best fits for the cyclobutyl resonances are shown in Figures 6 (a) and 7 (a). We generated several sets of internal coordinates for the cyclobutane rings corresponding to the folding

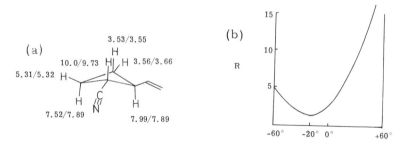

Figure 6. (a) Best fit of the observed/calculated cyclobutyl resonances in *trans*-1-cyano-2-vinyl cyclobutane; (b)Minimum agreement factors as a function of bending in the cyclobutane ring.

of the ring up to 60° either side of the average plane. Minimum agreement factors were obtained for these various

164

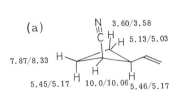

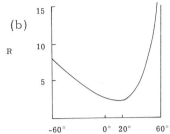

Figure 7. (a) Best fit of the observed/calculated cyclobutyl resonance in *cis*-1-cyano-2-vinyl cyclobutane; (b) Minimum agreement factors as a function of bending in the cyclobutane ring.

conformations and the results plotted as shown in Figures 6 (b) and 7 (b).

In order to use this approach we have assumed that the energy surface for the cyclobutane ring has but one minimum. If this is the case, then the graphs in Figures 6 (b) and 7 (b) specify that the best conformation for the cyclobutane ring in both cases is that in which the ring is folded 20 ± 5° from planarity. Furthermore, we used a coordinate system such that −20° of fold in the *trans* isomer corresponds to the conformation with equatorial cyano and vinyl groups; the +20° fold for the *cis* compound is that with an axial cyano and equatorial vinyl group. These conformations are in accord with previous determinations of the geometry of cyclobutane rings (14), and bode well for the use of the LIS experiment to assay the structure of molecules in solution.

We are presently trying to evaluate our assumption that only one conformer is present in the solution by computing shifts for two conformations and forming a weighted mean using appropriate mole fractions as weighting factors. This effort is not yet complete, but we have already seen indications that the improvement in fit will be marginal at best. We tentatively conclude that only one conformation of the cyclobutanes appears populated to the LIS experiment.

Thus far, we have described carefully constrained cyclic systems while omitting any discussion of the conformations of alicyclic chains. We are certain that non-rigid systems will require all of the external evidence one can gather in addition to the LIS simulations for a proper description. A case in point was provided by Faulkner and Stallard as part of their investigation of the chemical make-up of Pacific marine organisms. The California Sea Hare (*aplysia Californica*) contains the tetrahalo terpene alcohol shown here. The structure was demonstrated by

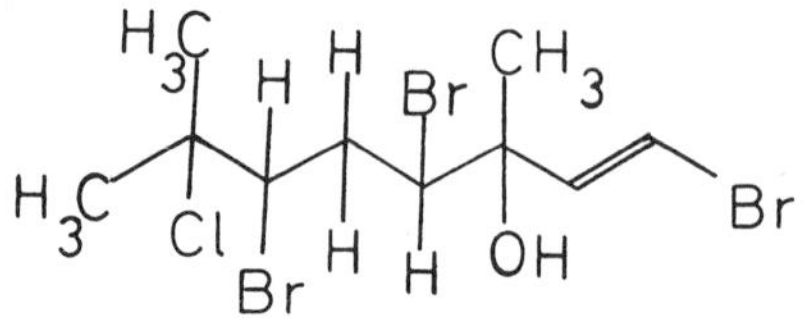

degradation and by synthesis, but the conformation and configuration was not. The olefin and three asymmetric centers introduce 16 stereochemical ambiguities. The *trans* coupling constant of the olefinic protons (J = 13 Hz) removes that ambiguity, and if we then consider only the relative configurations, the problem is reduced to deciding which of four molecules is the correct one. Conversion of the alcohol to an epoxide by base treatment is a chemical reaction which complements and simplifies the impending LIS experiment. The *cis* arrangement of the methyl and hydrogen on the epoxide ring were established by observing a 21% Nuclear Overhauser Enhancement. We emphasize that this much of the molecular description was done by Faulkner and Stallard without any recourse to the shift-reagent technique (15). The epoxide was treated with $Yb(DPM)_3$ to give the LIS indices shown in Figure 8. We

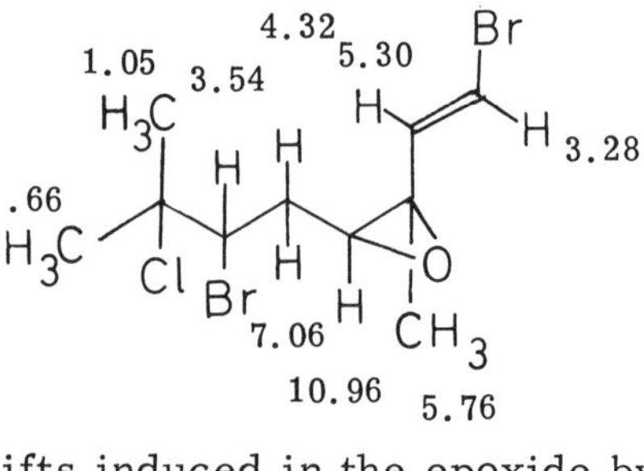

Figure 8. Shifts induced in the epoxide by $Yb(DPM)_3$.

used the two ring resonances in a PDIGM computation and found that even though there are only two observations, the ytterbium must still be located in the cross-hatched region for a reasonable agreement factor (Figure 9(a)). We then

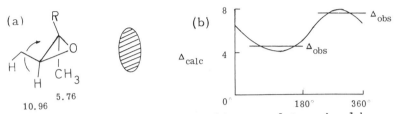

Figure 9. (a) Locations for ytterbium as determined by two ring resonances; (b) calculated shifts for the unlabeled hydrogen atom *vs.* dihedral angle.

provided a dummy hydrogen on the first carbon removed from the ring, rotated it in small increments for 360°, and computed an LIS value at each step. One of the graphs we used to summarize this operation is shown in Figure 9(b). Obviously, the many acceptable ytterbium locations gave graphs of differing magnitude, but all of them had a similar shape. We decided that the best geometry for the methylene group would be the one in which one of the hydrogens would have a calculated shift of about 7 ppm and the other about 4.5 ppm, with the additional restriction that the two hydrogens be separated by a tetrahedral angle. Simulation of the spectra with the fixed partial structure in Figure 10(a) localized the ytterbium position somewhat and let us perform the same computation for a hydrogen atom

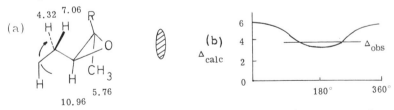

Figure 10. (a) Possible location for ytterbium using four hydrogen atoms; (b) calculated shifts for the unlabeled hydrogen atom *vs.* dihedral angle.

located on the second carbon. The calculated shift *vs.* conformation is shown in Figure 10(b). We have rotated the hydrogen atom in a clockwise sense, starting from the position shown in the Figure. Notice that LIS computation has reached an impasse. Since there are two locations for this atom which give reasonable calculated values, we cannot determine which conformation is the correct one. In the haste to apply LIS methods, it is easy to overlook the additional information at hand. The coupling constants in this molecule are readily determined by first-order analysis and provide the key to the rest of the experiment. The proper hydrogen conformation is the one which is *trans* to the resonance with index 4.32 (J (3.54, 4.32) = 8 Hz) and *gauche* to the other (J(7.06, 4.32) = 1 Hz). Figure 11 (a)

(a)

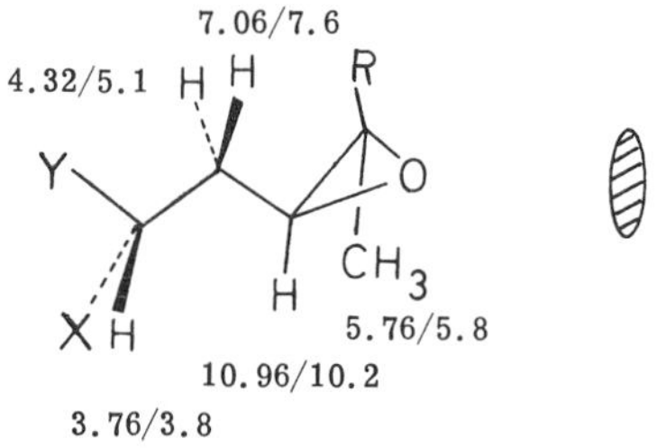

(b)

X = Br Y = $(CH_3)_2ClC$	OBS	X = $(CH_3)_2ClC$ Y = Br
.66	.66	3.68
.96	1.05	1.00
1.59		2.12

Figure 11. (a) The conformation of the side chain of the epoxide and observed/calculated values; (b) calculated shift for the two methyl groups at the end of the chain, and their experimental values.

shows the conformational assignments, the partial structure of which we are now certain, and the observed/calculated fit.

The task of assigning the configuration of the Br and alkyl groups (X and Y in the Figure), remains. The methyl groups give rise to proton nmr signals so we must examine their hypothetical LIS behavior. We did this by testing the three staggered rotamers for the geminal dimethyl group for both possible configurations. In the event that the alkyl group occupied the X position, we obtained the three LIS values to the left in Figure 11 (b), and when it occupied the Y position, we obtained the values at the right of the Figure. The observed values shown in the center of the Figure agree with the assignment X = alkyl, and not with the assignment Y = alkyl. Thus, we have determined the conformation and configuration of an acyclic moiety which can, in turn, be related to the terpene alcohol from aplysia to give the relative configurations at the three asymmetric centers in the molecule. It is tempting to offer this example as a structure determination of a conformationally mobile acyclic molecule, but for one fact. The coupling constants, invariant with added ytterbium, suggest the molecule is held predominantly in one conformation and that, for the purpose of our LIS analysis, it is rigid.

These examples, the xanthones, cyclobutanes, and epoxides are typical in our experience. The inherent resolution of the LIS experiment limited the accuracy of the results, but the topology of the substrates is well enough defined to encourage the continued use of the LIS experiment for studying conformational problems.

ACKNOWLEDGEMENT: This work was generously supported by the Robert A. Welch Foundation. One of us (MRW) expresses his thanks to the John Simon Guggenheim Memorial Foundation for a fellowship, and to the California Institute of Technology Division of Chemistry for providing a stimulating environment in which to pursue these studies.

References

1. Leading references for the literature through March 1973 are found in the preceeding paper.
2. The papers presented by Roberts, Davis, and Sweigart during this symposium are concerned with the use of a single L–S structure to interpret the LIS experiment.
3. R. E. Davis and M. R. Willcott, III, "Assessment of the Pseudocontact Model via Agreement Factor", in this Volume.
4. I. M. Armitage, L. D. Hall, A. G. Marshall, and L. G. Werbelow, J. Amer. Chem. Soc., 95, 1437 (1973), and this symposium.
5. N. S. Angerman, S. S. Danyluk, and T. A. Victor, J. Amer. Chem. Soc., 94, 7137 (1972).
6. C. D. Barry, A. C. T. North, J. A. Glasel, R. J. P. Williams, and A. V. Xavier, Nature, 232, 236 (1971).
7.a) J. Grandjean, Bull. Soc. Chim. Belges, 81, 513 (1972).
   b) N. H. Anderson, H. Uh, S. E. Smith, and P. G. M. Wuts, Chem. Commun., 956 (1972).
   c) C. Beauté, Z. W. Wolkowski, J. P. Merda, and D. Lelandais, Tetrahedron Letts., 2473 (1971).
   d) S. Farid and K.-H. Scholz, J. Org. Chem., 37, 481 (1972).
   e) P. Granger, M. M. Claudon, and J. F. Guinet, Tetrahedron Letts., 4167 (1971).
   f) W. Walter, R. F. Becker, and J. Thiem, Tetrahedron Letts., 1971 (1971).
   g) T. H. Sidall, III, Chem. Commun., 452 (1971).
   h) G. Montuado and P. Finnocchiaro, J. Org. Chem., 37, 3434 (1972).
8. J. J. Uebel and R. M. Wing, J. Amer. Chem. Soc., 94, 8910 (1972).
9. For a related LIS study of some flavones, see H. Okigawa, N. Kawano, W. Rahman, and M. M. Dahr, Tetrahedron Letts., 4125 (1972).
10. G. H. Stout, T. Shuh Lin, and I. Singh, Tetrahedron, 25, 1975 (1969).
11. P. H. Mazzochi, H. L. Ammon, and C. W. Jameson, Tetrahedron Letts., 573 (1973).
12. Z. W. Wolkowski, Tetrahedron Letts., 821 (1971).
13. The term "principal magnetic axis" is defined in our algorithm as the line connecting the lanthanide and

the origin of the coordinate system in which the substrate is described. We are concerned that this quantity is not the real magnetic axis. However, we find it convenient to use this unproved magnetic axis to describe the L–S structure in easily visualized terms, in spite of the potential error. Perhaps it is more properly called an effective magnetic axis.

14. The evidence for the folding of cyclobutane rings is summarized in J. B. Lambert and J. D. Roberts, J. Amer. Chem. Soc. , 87, 3884 (1965).

15. D. J. Faulkner and M. O. Stallard, Tetrahedron Letts. , in press.

THE STRUCTURE OF A CHOLESTEROL:
SHIFT REAGENT COMPLEX IN SOLUTION

C.D. Barry*, C.M. Dobson**, D.A. Sweigart***,
L.E. Ford*, and R.J.P. Williams**

*Department of Biophysics, Oxford, England
**Department of Inorganic Chemistry, Oxford, England
***Department of Chemistry, Swarthmore College,
   Swarthmore, Pa. 19081

Abstract: A study of the effect of shift reagents
$\overline{Ln(III)}(DPM)_3$, where Ln(III) is Eu(III) or Dy(III), and
a relaxation reagent Gd(III)(DPM)$_3$ on the ^{13}C and ^1H nmr
spectrum of cholesterol in chloroform has been
determined. These data have been used to determine the
structure of the Ln(III)(DPM)$_3$· cholesterol adduct.
The error limits of the structural analysis are
described.

Introduction

Cholesterol is a suitable compound for
testing our methods for determining structure
in solution by nmr spectroscopic probes. It is
made up of a rigid frame and a flexible chain.
It was used in an early structural study by
Hinckley[1] but our parallel experience in the
use of shift probes in aqueous media[2] led us to
suspect that the methods used in organic
solvents by Hinckley and subsequently by others
are unreliable. While in no way do we wish to
detract from Hinckley's contribution in
introducing shift reagents, there are several
grounds for doubt with regard to the present-day
uses of organic shift reagents in structural

173

studies. The problems can be seen by reference to the shift equation.

The lanthanides with fast electron relaxation times cause shifts of the resonance of nuclei according to the equation

$$\text{shift} = \text{constant} \left(\frac{3\cos^2\theta - 1}{r^3} \right)$$

for the case of <u>axial</u> symmetry; r is the distance of a nucleus from the metal ion, and θ the angle between the principal magnetic symmetry axis and the nucleus in question. If this equation is to be used it is necessary to prove that:

(a) the observed shifts are solely pseudo-contact in origin.
(b) the complexes have axial symmetry.
(c) there is no diamagnetic shift on complex formation.
(d) the shifts do not arise from an averaging over many different structures.

As we have shown[2,3] each of these assumptions can be examined by experiment: (a) and (b) may be tested by using more than one paramagnetic Ln(III)(DPM)$_3$ with fast electron relaxation times, and, preferably, choosing the different Ln(III) so that the shifts are of opposite sign. By using La(III)(DPM)$_3$ as a blank any diamagnetic shifts due to complex formation (c) are observed. A computer search, carried out in a particular way, can ensure that averaging is not important (d). A check on the structural analysis is provided by using relaxation probes. Using Gd(III) which has a slow electron relaxation time

$$\frac{1}{T_{iM}} = \text{constant} \left(\frac{1}{r^6} \right) \quad (i=1,2)$$

Taking ratios of these effects on different nuclei in the molecule enables the constants to be eliminated from the equations and so there is a direct check on r independent of θ. We shall show in this paper the power of our procedure as applied to the adducts of organic shift reagents.

Experimental

Cholesterol was obtained from Sigma Chemical Co. Ltd. The Ln(DPM)$_3$ chelates were made by standard procedures and gave satisfactory analyses. All pmr spectra were run in CDCl$_3$ with TMS as internal standard for protons, DMSO for ^{13}C. The steroid concentration ranged from 0.1 to 0.25 M without affecting the shift ratios, which were also independent of Ln(DPM)$_3$ concentration. The temperature was held at 18°C.

PMR spectra were run using a 60 MHz JEOLCO spectrometer and a Brüker 270 MHz instrument employing an Oxford Instrument Company magnet and Nicolet 1085 computer. ^{13}C spectra were run on a Brüker HX90 operating at 26.36 MHz, and using 10 mm tubes.

Results

In the description and discussion of our experiments we shall use the numbering scheme of Fig. 1.

The proton and carbon shifts (ratioed to C$_4$ for cholesterol are given in Table I. The ratios were not changed by more than $\pm$ 0.01

when Eu(III) was replaced by Dy(III). In
keeping with this result, there was no
measurable shift on adding La(III)(DPM)$_3$ to
cholesterol. Thus we can assume the shifts as
pseudo-contact in origin and that the ligand
field has effective axial symmetry.

In the spectrum of cholesterol, irradiation
of the H$_{26}$ resonance allowed the assignment of
the position of H$_{23}$ and H$_{24}$. Equally irradiation at this position caused collapse of the
H$_{26}$ resonance. Irradiation at the H$_{10}$ resonance
allowed the positions of the four protons H$_{13}$,
H$_{14}$, H$_7$, H$_8$ to be determined by a difference
spectrum. To distinguish these protons,
irradiation at the low field position caused
changes only at the H$_{10}$ resonance, thus allowing
assignment to H$_{7,8}$ (in agreement with
Hinckley[1]). Similarly, the larger peak in the
difference spectrum can be assigned to the axial
H$_{13}$, (larger coupling constant) the smaller peak
to H$_{14}$. The shifts of peaks assigned to these
resonances in the shifted spectra can be
extrapolated back to these positions. The
larger coupling observed for H$_7$ as compared to
H$_8$ and H$_{13}$ to H$_{14}$ in these spectra confirms that
H$_7$ and H$_{13}$ are the axial protons. Other
resonances were assigned after the initial
structure determination on the basis of the
correlation of observed shifts with those predicted from our structure. The assignments[4]
are shown in Fig. 2.

The ^{13}C resonance assignments were taken from
Roberts[5] and were all consistent with the shift
data.

Knowing the assignments, the observed shift
ratios for the different nuclei can be used in
the computation of the structure of cholesterol-

Ln(III)(DPM)$_3$ adduct. In order to carry out this analysis we need an initial frame for a limited part of the molecule. There is no structural study of cholesterol in the solid state. However, a good construction of its geometry can be obtained in the following way.[6] The X-ray structure of a known steroid[7] was used for the carbon atoms in the fused ring system. The -OH was placed in position at a tetrahedral angle. The protons were generated so as to be 1.07Å from each carbon at tetrahedral angles (except H$_{26}$). Table II gives the coordinates of the frame.

Conformational Analysis Of The Structure

The analysis starts on the basis of part of the rigid frame for cholesterol described above. To the co-ordinates of carbons 6, 12, 20, 22 and 25 were added those of protons 8, 10 and 26. The nmr shift data for these nuclei was used by the computer program to establish the position of the metal and its symmetry axis (assuming axial symmetry). The analysis proceeded in terms of the polar co-ordinates R, ϕ and ψ of the metal with respect to carbon 9. The symmetry axis was given the similar variables α and β as shown in Fig. 3. The computer output consisted of values of R, ϕ, ψ, α, β consistent with the nmr data. Eventually R was searched over the range 3-5Å at 0.1Å intervals, and the angles were scanned over all ranges at 2° or 4° intervals. No atoms corresponding to the DPM ligands were included. A preliminary coarse analysis with nmr tolerances set high gave acceptable solutions only when 320°>ψ>290° and 20°<ϕ<35° with R values in the range 3.8-4.2Å. A fine scan was made for R = 3.8 - 4.2Å in steps of 0.1Å, ϕ = 20 - 36° by 2°, ψ = 290-322° by 4°.

This showed $\alpha > \phi + 10°$. Tolerances were then
reduced to those given in Table I to limit the
number of solutions to several hundred. However, all tolerances were carefully examined to
ensure that no single value determined the
outcome. After analysis of all output, the
values given in Table III were obtained.

The nature of the "structure" found can now
be discussed by considering the number of shift
solutions at a particular value of each parameter, allowing all other parameters to vary
over the range of permissible solutions. Within
the range of solutions accepted, the symmetry
axis and the M-O bond coincide to $\pm 1°$. For
most solutions $\beta = \psi$, though it should be noted
that there were some fairly good solutions
having β different from ψ by $\pm 4°$ but these were
only 5% of the total number of solutions.

For R outside the range 3.8 - 4.0Å few
solutions could be found. The most solutions
were at R = 3.9Å, and a typical very good
solution is represented in Fig. 4. (R = 3.9Å,
$\phi = 26°$, $\psi = 298°$, $\alpha = 40°$, $\beta = 298°$). The M-O
bond length is 2.69Å here, and the MOC angle
140.5°. The bond length is very reasonable.
The angle is somewhat higher than anticipated
but it could be attributed to a "steric effect".
The greater the MOC angle (smaller ϕ) the more
the molecule of cholesterol is directed away
from the DPM ligands, and this probably reduces
steric interactions.

As shown in Fig. 5 the M-O bond is rotated
by 25-30° with respect to the approximate plane
of symmetry (yz plane) through the first fused
ring (A). The rotation is towards the positive
X axis, i.e. the metal X coordinate is positive.
Inspection of a model shows that this position

probably involves less steric interaction than the straight on position ($\psi = 270°$). A model does not make it immediately obvious why the molecule does not rotate the other way ($\psi < 270°$) but we note that cholesterol is not symmetric through the ring system and that this could easily impart the necessary energy difference.

Errors

The observed shifts can be fitted to within about 5% except for a few atoms with small shifts. We can be certain of all observed shift data to within 10% but for the calculated shifts we must add in any error in the input structure. For atoms close to the metal, an error of a few tenths of one Angstrom can be significant, as can angle factors. Since cholesterol was synthesized by joining various parts of other molecules an additional 5 or 10% error can be introduced to allow for structure faults, particularly for protons. Thus, in a comparison of this kind any calculated shift within 15% of the observed is quite satisfactory. Table I shows that the average % error is 9.3%.

Relaxation

By shifting the resonances using $Eu(DPM)_3$, it is possible to separate many individual peaks at particular concentrations of $Eu(DPM)_3$. Addition of $Gd(DPM)_3$ causes broadening of these resonances, and this broadening is proportional to $1/r^6$ where r is the distance of the proton from the metal. This broadening shows clearly that the distances of these protons from the metal ion are in the order:

$H_{27} < H_{10} < H_8 < H_{14} \approx H_7 < H_{13} < H_{24} < H_{26} < H_{23} < H_{15-17}$ $< H_{36-38} <$ other methyl groups.

179

We have not attempted to quantify these (T_2)
data because of the high degree of coupling in
these systems. The order is in agreement with
the structure obtained from the shift data. It
quite clearly demonstrates that the metal
complex lies on the underside of the molecule
away from the methyl group H_{15-17}, and towards
the H_{26} proton. In order to quantify these
distances spin-lattice relaxation times (T_1)
were measured by means of a $180°$-τ-$90°$ pulse
method. The coupling of the protons is not
important. Table IV summarises the data on
several protons. Again, the qualitative agreement is excellent, and as any error in the ring
pucker or bond length would affect the data on
H_{10} in particular, quantitative agreement is
satisfactory. Note, that a positioning of the
metal with $\psi \approx 90°$, would make the distance to
$H_{15-17} < H_{26}$, $H_{13} < H_{14}$, and $H_7 < H_8$. This emphasises
the importance of the angular factor, for in
fact the shift magnitudes are in this order,
but experimental distances are not.

A conclusion from the measurement of
diamagnetic T_1 (in the absence of $Gd(DPM)_3$)
concerns the rigidity of the molecule (Table V).
The data on the ring system can be explained on
the basis of a dipolar relaxation process, and
rapid rotation of the methyl groups, the longer
T_1 for the chain CH_3 groups implies a less
rigid structure for this part of the molecule,
in agreement with ^{13}C measurements. For this
reason (and because the observed shifts, even
with $Dy(DPM)_3$, were small) no attempt was made
to fit this part of the molecule to a structure.
However, it is possible to state that there is
no folding back of the molecule.

Discussion

The structure that we have obtained differs significantly from that put forward previously[1] although there is little difference in the experimental data. The previous structure showed that the metal is directed towards the C_4 methyl group (i.e. y>0, $\psi \approx 90°$), because of steric interaction with H_{10} if $\psi \approx 270°$. The relaxation data has at once shown that this is not correct.

This conclusion is also abundantly clear from our shift data. There are no reasonable solutions for y>0. Steric interaction with the C_4 methyl appears much more likely than with H_{10}. Indeed, for $\psi \approx 90°$ this interaction is very probable. In addition, in earlier work it was decided to ignore the $\cos^2 \theta$ factor when considering the shift data, and so it was deduced that $\psi \approx 90°$ from a strict r^{-3} comparison from the shift of H_{26} compared to H_{15-17}. Our results show that the angle factor is important, and there are numerous examples of the nearer atom of two giving the smaller shift (Table VI). For example both C_O and H_{26} are $6.3\overset{\circ}{A}$ from the metal yet C_O has three times the shift of H_{26}. A model shows clearly that H_{26} is nearer the edge of the shift "cone" than C_O. Some sample shifts vs calculated distances are given in Table VI (from the solution R = $3.9\overset{\circ}{A}$, ϕ = 28°, ψ = 294°, α = 42°, β = 294°).

Thus neglect of the $\cos^2 \theta$ factor is inappropriate. It is frequently assumed that log (shift) = -3 log R + constant. Fig. 6 shows some of our data on such a plot. Most of the shifts approximate to a line of slope -3 but there is a lot of scatter.

181

We note that the structure which we have found can be obtained in different ways from the addition of cholesterol to one mole of the Ln(DPM)$_3$ complex. The fact that axial symmetry is observed follows if exchange between these sites is fast. It would also be the case that fast rotation about the Ln(III)- O bond (calculated to be the symmetry axis) will give effective axial symmetry.[8] We cannot distinguish between any set of molecular motions which will give rise to such symmetry.

We conclude from the arguments above, that structure   determinations in solution are possible, but great care must be taken to extract the input data from the experiments and proper computational searches must be made. If confidence is to be placed in the structure , it must be confirmed by relaxation studies.

Acknowledgement

D.A.S. acknowledges financial support in the form of a NATO postdoctoral fellowship.

References

1. C.C. Hinckley, J. Amer. Chem. Soc., 91, 5160 (1969).
2. C.D. Barry, J.A. Glasel, A.C.T. North, R.J.P. Williams and A.V. Xavier, Nature, 232, 236 (1971).
3. B. Bleaney, C.M. Dobson, B.A. Levine, R.B. Martin, R.J.P. Williams and A.V. Xavier, Chem. Commun., 791 (1972).
4. Varian catalog No. 363.
5. H.J. Reich, M.J. Autelat, M.T. Messe, F.J. Weigert and J.D. Roberts, J. Amer. Chem. Soc., 91, 7445 (1969).
6. P. Sadler, D. Phil. Thesis, p. 66, Oxford (1972).
7. A. Cooper et al., Acta Cryst., B24, 811 (1968).
8. J.H. Briggs, G.P. Moss, E.W. Randall and K.D. Sales, Chem. Commun., 1180 (1972).
9. I.D. Campbell, C.M. Dobson, R.J.P. Williams and A.V. Xavier, to be published.

Table I. Shift Data for Cholesterol

Atom	Observed shift[a]	Calculated	Error[b] %	Tolerance used
C 0	132	150	13.5	32
H 1	89	107	20	30
H 2	104	120	15	25
C 3	125	117	-6.5	22
C 4	100	100	-	-
C 5	114	138	21	33
C 6	314	318	1	20[c]
H 7	296	270	-9	50
H 8	246	235	-4.5	100
C 9	993	635	-36	500
H 10	433	425	2	150
C 12	314	312	-1	20[c]
H 13	296	285	-4	100
H 14	260	262	1	60
H 15-17	80	94	17	20
C 18	57	63	10.5	6[c]
H 19/21	49	60[d]	22	20
C 20	40	41	2	4
C 22	40	38	-5	6
C 25	85	62	-27	50
H 26	50	44.	-12	25
H 27	1604	2200	40	800

Table I. Shift Data for Cholesterol(Contd)

Atom	Observed	Calculated shift[a]	Error[b] %	Tolerance used
C 28	43	48	12	7
C 31	-	30	-	-
C 34	25	24	-4	4
C 35	21	21	0	3
H 36-38	19	20	5	3
C 39	25	26	3	3
C 41	18	16	-14	5
C 44	-	12	-	-
C 47	18	16	-14	5
C 48[e]	11	-	-	-
C 49[e]	11	-	-	-
H 50-52[e]	9	-	-	-
C 53[e]	14	-	-	-
H 58-60[e]	3	-	-	-

[a] Relative to C 4 = 100

[b] 100 (shift(calc) - shift(obs)) / shift(obs).
Average Error =$(\Sigma|\text{Error}|)/26 = 9.3\%$ (H 27 and
C 9 not included).

[c] Tolerance can be increased beyond these limits
without affecting the range in which solutions
are obtained.

[d] Calculated assuming H 19.

[e] This part of the molecule was not computed.

Table II. Coordinates[a] of the Fused Rings
of Cholesterol

Atom	x	y	z	Atom	x	y	z
C0	-122	81	-202	C25	224	40	-294
H1	-208	136	-235	H26	302	-4	-225
H2	-125	-16	-245	H27	69	67	168
C3	5	156	-256	C28	-98	235	-472
C4	15	304	-213	H29	-65	336	-449
C5	125	85	-198	H30	-193	216	-426
C6	126	71	-47	C31	-111	221	-622
H7	132	178	-4	H32	-168	304	-660
H8	212	14	-17	H33	-163	129	-642
C9	0	0	0	C34	20	218	-697
H10	0	-101	-35	C35	82	360	-699
O11	0	0	143	H36	20	427	-637
C12	-129	67	-50	H37	184	356	-658
H13	-137	164	-5	H38	84	398	-802
H14	-214	6	-22	C39	115	117	-630
H15	-12	313	-108	H40	66	20	-619
H16	120	340	-228	C41	235	108	-724
H17	-61	365	-271	H42	305	188	-706
C18	7	137	-410	H43	284	14	-714
H19	-36	40	-422	C44	170	124	-866
C20	135	143	-480	H45	221	199	-922
H21	178	240	-471	H46	174	28	-919
C22	229	48	-416	C47	25	164	-844
H23	327	80	-445	H49	-41	79	-854
H24	209	-50	-455				

[a] Units are 0.01Å.

Table III.   Solutions for $Ln(DPM)_3 \cdot$ Cholesterol

Maximum variation	Probable variation
$\phi = 22\text{-}32°$	$\phi = 24\text{-}30°$
$\psi = 292\text{-}304°$	$\psi = 294\text{-}302°$
$\alpha = \phi + 15° \pm 1°$	$\alpha = \phi + 15° \pm 1°$
$\beta = \psi \pm 4°$	$\beta = \psi$
$R = 3.8 - 4.0\text{Å}$	$R = 3.8 - 4.0\text{Å}.$

Table IV.   Relaxation Data for
$Ln(DPM)_3 \cdot$ Cholesterol

Atom	Ratio of $1/T_1P$ to H_{26}	Distance assuming $H_{26}=6.34\text{Å}$	Structure distance
H_{10}	4.8 ± 2	4.9 ± 0.5	3.9
H_{26}	1.00	6.34	6.34
H_{36-38}	<0.08	>9.9	11.85
H_{15-17}	0.60 ± 0.20	6.85 ± 0.3	6.87

Table V.   Relaxation Data in Absence of Ln(DPM)$_3$

Ring system	Range of T$_1$
-CH	0.85 ± 0.1
-CH$_2$	0.40 ± 0.1
-CH$_3$	0.62 ± 0.05
Chain -CH$_3$	1.0 ± 0.1

Table VI.   Observed Shifts and Calculated
Distances for Some Atoms.

Atom	Observed Shift	r (Å)
CO	132	6.31
H2	104	6.40
H26	50	6.34
C25	85	6.88
C3	125	6.85
H7	296	4.94
H13	296	5.26
H8	246	4.27
H14	260	4.97

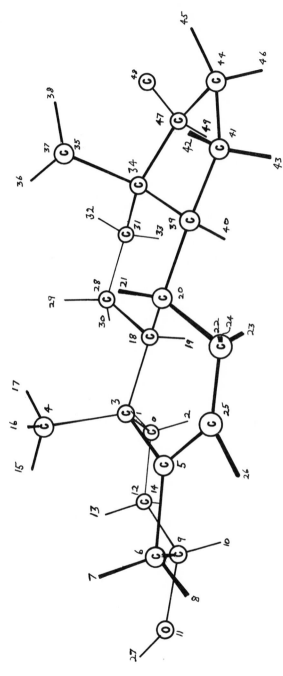

1. Numbering scheme for cholesterol.

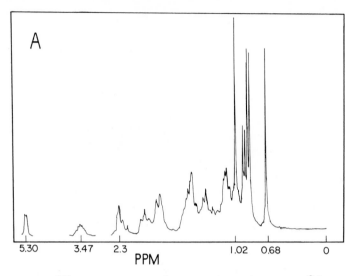

2. A. 270 MHz normal pmr spectrum, 18°C, of
    cholesterol.

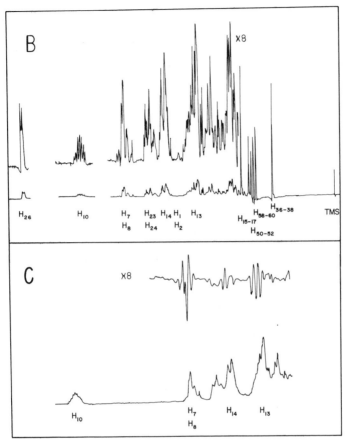

2. B. Convolution difference spectrum[9], with assignments, of A.

C. Irradiation of H_{10} resonance. Lower trace, normal spectrum; Upper trace, difference between spectra with and without irradiation.

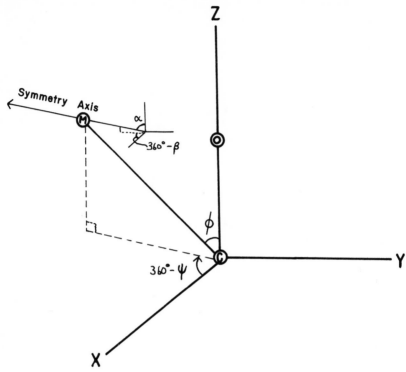

3. Parameters defining metal position and symmetry axis.

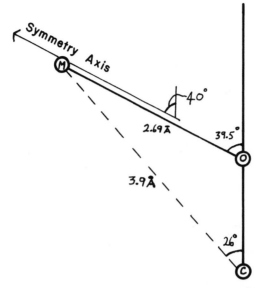

4. A typical solution, showing position of
    metal and symmetry axis.

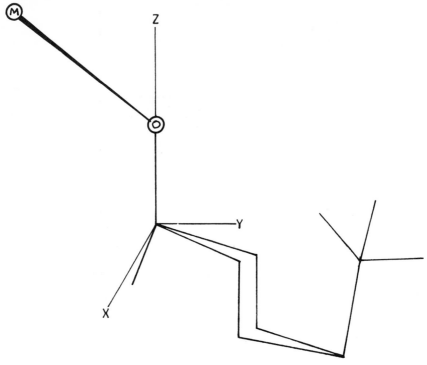

5. A typical solution showing position of metal relative to first fused ring of cholesterol.

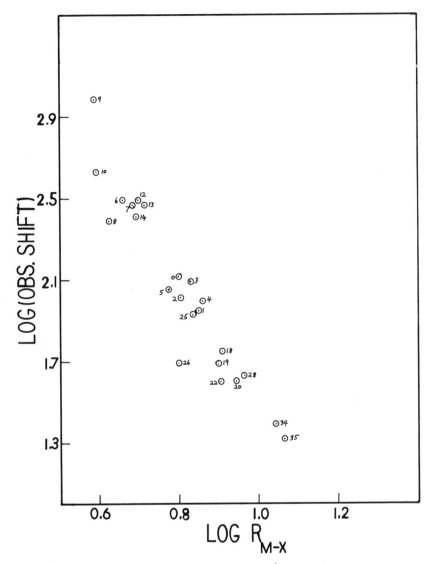

6. Plot of log (obs. shift) vs. log R_{M-X}
($X = H$ or C).

Studies of Lanthanide Shift Reagents at Queen Mary College.

By

J.M.Briggs, F.A.Hart, G.P.Moss, E.W.Randall, K.D.Sales, and M.L.Staniforth. (Department of Chemistry, Queen Mary College, Mile End Road, London E1 4NS.)

Introduction

A major problem in the complete interpretation of [1]H n.m.r. spectra arises from the small range of chemical shifts observed for many signals. This frequently results in the superposition of signals and in second order spectra. Although the use of spectrometers operating at 220 MHz[1] (or 300 MHz) may help to resolve most peaks and to simplify the spectra, there is an alternative approach, viz.,to run the spectrum of the sample in the presence of a paramagnetic complex. If certain conditions involving the relaxation times are obeyed,[2] the magnetic properties of the complex may induce shifts in the n.m.r. spectrum of the substrate being examined.

Some of the earliest n.m.r. work on paramagnetic compounds used stable complexes, coordinatively saturated, of transition metals, e.g., Ni^{2+}, Co^{2+}, Mn^{2+} and Cu^{2+}, and examined the n.m.r. spectra of the ligands. If a substrate molecule with a polar group, such as an alcohol or amine is present it also may have donor capacity. Replacement of a ligand and coordination of the substrate with the metal induces shifts in the substrate spectrum by one or

197

both of the pseudo-contact and electron delocalisation mechanisms. A serious disadvantage of this technique is that broadening of the signals often accompanies the shifts.

Paramagnetic lanthanide ions also induce shifts which in some cases are predominantly of a pseudo-contact nature. The earliest work[3,4] using these ions examined the n.m.r. spectra of the ligands, rather than those of a substrate. However, in 1969 Hinckley[5] showed that shifts of up to 3.47 p.p.m. were observed in the spectrum of cholesterol in the presence of Eu(tmhd)$_3$py$_2$ {tmhd = 2,2,6, 6- tetramethyl heptane-3,5-dionato ligand (i.e., ^{t}Bu CO.CH. CO ^{t}Bu); py = pyridine}. The suggestion that these shifts were predominantly pseudo-contact in nature was supported by the demonstration of a rough r^{-3} relationship. Two significant features of the spectra illustrated by Hinckley are that there was little broadening of the signals, and that only one set of signals[2] was observed for cholesterol. The latter observation implies that there is a rapid exchange between the free substrate molecules and those that are coordinated with the europium complex.

Within a short period several laboratories developed Hinckley's technique to yield numerous applications in organic (and inorganic) chemistry. Sanders and Williams[6] showed that the coordinatively unsaturated complex Eu(tm hd)$_3$ gave larger shifts than Eu(tmhd)$_3$py$_2$ did, and that substrates with a range of polar functional groups could be examined in this way. Following our previous studies,[4] we showed[7] that Pr(tmhd)$_3$ induced even larger shifts which were to high field, i.e., in a direction opposite to that produced by europium. In some cases complexes of the fluorinated ligand, fod (= $6,6,7,7,8,8,8$-heptafluoro-2,2-dimethyl-3,5-octanedione) give larger shifts[8] than $M(tmdh)_3$ does, owing to its greater solubility in non-polar solvents and better complexing ability with weak ligands such as ethers. With complexes having an optically active ligand, such as 3-(tert-butyl hydroxmethylene)-d-camphorato, enantiomeric substrates show different induced shifts.[9] In aqueous solution the free cation may be used to induce shifts if strong ligands such as carboxylates[10,11] or phosphates[12,13] are present.

Several hundred papers have appeared[14] describing the application of lanthanide-induced shifts. In most cases the aim has been to obtain first order spectra, and only relatively few attempts have been made to derive stereo-chemical information from the shift data. The main emphasis of our studies has been directed to an under-standing of the fundamental processes involved in lanthan-ide-induced shifts. In order to simplify the geometrical problems associated with the calculation of the pseudo-contact effects, many of our studies have been conducted with the relatively rigid alcohol borneol as substrate (see Fig.1).

Most of the lanthanides have been examined as possible shift reagents.[4,15] In general europium and praseodymium have proved most suitable for [1]H n.m.r. studies: they minimise the broadening effect, and are complementary since they shift signals in opposite directions. Broadening at low lanthanide concentration is undesirable since it prevents the resolution of small couplings which may be otherwise useful for assignments. It is more pronounced with praseodymium than europium (see Fig.2). At higher concentrations of lanthanide, broadening becomes a more serious problem, and leads to eventual loss of the signals. On the other hand, since the broadening is proportional to r^{-6}, this effect may in turn be used to yield structural information.[10,12]

Structure and Stoichiometry.

X-ray crystallographic studies of a number of lanthanide complexes show that the reagent $Pr(tmhd)_3$ is present in the solid state as a dimer[16] with a 7-coordinate metal ion. Because of the smaller ionic radius of erbium, $Er(tmhd)_3$. exists in the crystal as a monomer with 6-coordinate triangular prismatic coordination.[17] Several adducts have also been examined, such as the 7-coordinate $Lu(tmhd)_3$. C_6H_7N[18] and $Eu(tmhd)_3.Me_2C(CH_2)_2SO$.[19] 8-coordinate complexes are also known in the solid state, for example $Eu(tmhd)_3py_2$.[20] The 7-coordination polyhedron may be approximately described as a capped triangular prism. The substrate ligand is one of the four forming the capped face (see Fig.3). The 8-coordinate examples have an approximate square antiprismatic structure. It is not possible to predict from these X-ray studies the nature of the species in solution of course, or even to predict the stoichiometry of the interaction between substrate and complexes.

The n.m.r. spectra of freshly sublimed $M(tmhd)_3$ obtained by use of vacuum line techniques show only two peaks (M = Pr, Eu)- a high field one due to the tert-butyl groups and a weak low-field peak from the vinyl protons.[21]

However, without stringent precautions to exclude moisture, an additional methyl signal is present. Although this peak has been assigned[22] to the dimer, it clearly originates from a hydrated species since its intensity increases with time if ingress of moisture is allowed (see Fig.4). Further-more the molecular weight of the species in solution has been shown by osmometry to correspond to the monomer, M(tmhd)$_3$ (M =La,Pr,Eu,Lu).[21] Ebulliometry in the presence of excess borneol or β-naphthoquinoline showed that 0.82 - 0.85 molar proportions of the substrate were inactive in producing any temperature rise; presumably a 1:1 adduct had been formed. More definite evidence for this stoichio-metry was obtained[21] by the application of Job's method to an investigation of the effect of added substrate on the intensity of the f-f absorption bands of Ln(tmhd)$_3$ comp-lexes. The maxima of these plots correspond to a 1:1 stoichiometry. Other work with the fod complexes however suggests that, with the smaller steric requirements of the perfluoro-n-propyl group compared with a tert-butyl group, a 2:1 adduct (i.e., two molecules of substrate to one of complex) is more readily formed.[23] The general conclusion from our results is that a 1:1 complex is formed between M(tmhd)$_3$ and the substrate.

Theory of anisotropically induced shifts.

The pseudo-contact contribution to the induced shift of a given nucleus in the complex should be expressed by the equation (1).

$$\frac{\Delta H}{H_0} = K_1 \frac{(3\cos^2\chi-1)}{r^3} + K_2 \frac{\sin^2\chi \, \cos2\psi}{r^3} \qquad (1)$$

The spherical polar coordinates of the nucleus are given by r, χ, and ψ, where the origin is the paramagnetic centre and the polar axis is one of the principal magnetic susceptibility axes. The constants K_1 and K_2 for a given complex involve the principal magnetic susceptibility components (χ_1, χ_2, χ_3).[24]

A number of studies have analysed the lanthanide induced shifts on the assumption that there is axial symmetry in the susceptibility tensor, and that the principal axis is colinear with the lanthanide-ligand bond,[25,27] or is defined by the ligand.[12,28] With these assumptions equation (1) simplifies to equation (2) where θ is now the polar angle referred to the new principal axis.

$$\frac{\Delta H}{H_o} = K\frac{(\cos^2\theta - 1)}{r^3} \tag{2}$$

Despite the fact that equation (2) appears to rationalise many of the observed shifts, the assumptions leading to this equation do not seem to be applicable to the cases considered since the X-ray studies reveal too low a symmetry. We have shown,[29] however, than an equation similar to (2) may be valid for other reasons.

We transform coordinates of equation (1) to a new set of axes in which the z-axis is a rotation axis, and is related to the old z-axis by the Euler angles (β, α, θ) (see Fig.5). The rotation axis must pass through the lanthanide. The polar coordinates of a nucleus in the new frame are $(r, \theta, \phi,)$. If this nucleus can adopt positions such that $\phi = \phi_1$, $\phi = \phi_1 + 2\pi/n$, $\phi = \phi_1 + 4\pi/n$, etc., where ϕ_1 is an arbitrary angle and n is a positive integer, and the ligand can move within $\pm\gamma$ of the mean angle, then the observed shift can be described by equation (3).

$$\left\langle\frac{\Delta H}{H_o}\right\rangle = \frac{\displaystyle\sum_{i=0}^{n-1}\int_{\phi_1 +2\pi/n-\gamma}^{\phi_1 + 2i\pi/n+\gamma}\left(\frac{\Delta H}{H_o}\right)d\phi}{\displaystyle\sum_{i=0}^{n-1}\int_{\phi_1 +2i\pi/n-\gamma}^{\phi_1 +2\pi/n+\gamma}d\phi} \tag{3}$$

If $n \geqslant 3$, and if there are equal populations for the conformers, then we derive equation (4):

$$\frac{\Delta H}{H_o} = \tfrac{1}{2}\{K_1(\cos^2\alpha-1) + K_2\sin^2\alpha \cos2\beta\}\frac{(3\cos^2\theta-1)}{r^3}$$

$$= K_3\frac{(3\cos^2\theta-1)}{r^3} \tag{4}$$

The case of $\gamma = \pi/n$ corresponds to free rotation of the ligand. Equation (4) is similar to equation (2), except that K_3 is now a function of α and β as well as the magnetic susceptibility components. However, for a given 1:1 complex α and β are fixed and K_3 is therefore a constant. It is now apparent that equation (4) should be valid if the substrate ligand undergoes free rotation about an axis passing through the lanthanide ion, or if it forms three (or more) interconverting rotamers which are equally populated.

Applications.

We have developed computer programs which make use of equation (4) to calculate an optimised position for the lanthanide atom based on a comparison between the observed and the calculated shifts. First, K_3 is determined by equation (5),

$$K_3 = \frac{\sum\limits_{i=1}^{n} (\Delta H_i^{obs})^2}{\sum\limits_{i=1}^{n} \Delta H_i^{obs} \, \Delta A_i^{calc}} \tag{5}$$

where n nuclei are examined, the i-th nucleus has an observed shift of ΔH_i^{obs}, and $\Delta A_i^{calc} = (3\cos^2\theta_i - 1)r_i^{-3}$. The R.M.S. deviation (R) between the observed and calculated shifts is given by equation (6),

$$R^2 = \sum\limits_{i=1}^{i=n} (\Delta H_i^{obs} - K_3 \Delta A_i^{calc})^2 \tag{6}$$

The program then progressively moves the lanthanide atom so as to minimise the value of R. Because the shape of the surface defined by R is unknown, it was decided not to use a general optimising method but instead to crank the position of the lanthanide ion along the x axis until a minimum value of R is located. The process is then repeated along the y and z axes and the correction vector of these three movements. After each cycle of four optimisations the size of the crank is reduced and the process is repeated as many times as required. A graphic picture of such a calculation is given in Fig. 6, where by intention the starting position is chosen to be chemically ridiculous, but in which the large movements demonstrate how the program operates.

An excellent agreement was found between the observed proton shifts for borneol and the values calculated from

equations (4) and (5) (Table 1). The final value of R
was 0.66 p.p.m. To examine whether any contact effects
were present, the closest hydrogen atom (2-exo see Fig.1),
which should be most affected in this way, was removed
from the optimisation. This had virtually no effect on
the optimised lanthanide position and gave a value of R
of 0.68 p.p.m., which is slightly worse. The ambiguity
over the assignment of the gem-dimethyl group (8 and 9)
was checked by reversing their assignments. A small but
distinct improvement was observed (R = 0.62 p.p.m.).

An optimisation using all ten observed[26] ^{13}C shifts
gave R = 1.26 p.p.m. (Table 1). The agreement is slightly
worse than for the proton case partly because of the
inherently larger errors in our measurements of ^{13}C shifts.
The dominance of the pseudo-contact interaction is confirmed by the excellent agreement between (i) the calculated and observed shifts for each of the ^{1}H and ^{13}C cases;
and (ii) the values of K_3 derived independently from the
^{1}H and ^{13}C data.

Removal of C(2) from the optimisation procedure for
^{13}C improved the fit to R = 1.06 p.p.m. Thus in contrast
to the proton case removal of the nucleus closest to the
lanthanide gave a significant improvement. This may be
due either to some significant contact contribution to the
C(2) shift, or to a breakdown of the point dipole approximation in the usual derivation of the pseudo-contact
formula (equation 2).

Additional errors which contribute to R may arise
from several sources, for example: the coordinates used

were based on a Dreiding model; for each methyl group the
proton position was taken as being the mid-point of the
three hydrogen atoms; and a single preferred rotamer
about the bond between the oxygen and C-1 was assumed.

One consequence of equation (4) is that when θ is
greater than $54° 44'$, the sign of the induced shift is
reversed. One of the first examples of this effect was
observed[30] with 2,6-diisopropyl acetanilide (Fig.7). With
the E isomer shown, the methyl group trans to the acetyl
group shows a negative shift. On the assumption that the
planes of the amide unit and of the aromatic ring are
perpendicular (e.g., see ref. 31), and that the isopropyl
methine hydrogen is in the plane of the ring, the calculated values for this conformer show a close fit with the
observed values. It is of note that the position of the
europium is as expected for coordination with the amide
oxygen lone pair (see Table 2.). Calculations using only
one of the isopropyl groups and the acetyl methyl gave a
negligible change in the europium position.

Similar calculations may be made on shift data from
aqueous solutions. The carboxylate shown in Fig.8 was
examined using $Pr (ClO_4)_3$ in D_2O. If we assume that the
carboxylate acts as a bidentate ligand and that free
rotation is possible about the bond from C-4 to the carboxyl carbon, our calculations predict that the optimum
praseodymium position is within 0.18 Å of the value
estimated from X-ray data for the nitrate (see Table 3).
The presence of two sp^2 hybridised carbon atoms in the
terpenoid model compound (Fig.9) presents an interesting
problem as to the preferred conformation that the A-ring

206

will adopt. Three conformers can be formed readily from
Dreiding models, but each has serious 1,3-interactions
(chair 4β-Me to 10-Me; 1,4α-boat, 4α-Me to 1α-H; 1,4β-boat
4β-Me to 1β-H). However a model with C-2, 3, 4, and 5
planar minimises these interactions. Calculations of
lanthanide induced shifts based on these models (Table 4)
showed that only the last conformer gave results which
were consistent both with the observed shifts and with a
reasonable position for the lanthanide atom (a Pr-O
distance of over 3 Å is chemically unlikely). These calculations however do not take into account any conformational
mobility, whereas the observed shifts are a time-average
of all conformers present. The most that can be said is
that the calculation is consistent with the presence of
one conformer.

An important use of this type of calculation is for
the assignment of ^{13}C n.m.r. signals. This approach was
first checked with borneol (see Fig.10 and Table 1).
Previous work[32] had assigned the signals from the norborn-
ane skeleton. By use of the lanthanide-induced shifts[26]
not only were these partial assignments confirmed but the
three methyl groups were identified as well.[†] In the case
of cedrol, initially only three signals could be identified
readily,[33] and which C-6 was not certain. The assignments
shown in Fig.10 are based on calculations using the
results with three lanthanides (Eu,Pr, Ho). The subtle

† Footnote: The effects of "shiftless" relaxation
reagents on the T_1's of the ^{13}C nuclei of borneol confirm
the whole assignment (G.C.Levy, personal communication.)

steric effects with this tertiary alcohol are shown by
Table 5. Whereas C-7 and the 6S-methyl are shifted the
same amount by europium, with holmium they are quite
different. In contrast C-9 and the 6R-methyl are shifted
by the same amount with holmium and are quite different
with europium. These differences can be rationalised by
a rotation about the O-C bond of 15° from one complex to
another.

Studies of lanthanide induced shifts may not answer
all problems of the type outlined here. One limitation
arises from the presence of more than one polar group.
This may result either in a complex having a bidentate
substrate ligand, or in two complexes with the substrate
acting as a monodentate ligand in each case but coordinated
differently.[34] Through-bond contact effects are more
likely to be a problem where the ligand atom is part of
a π -system. This also seems to be a problem with
aliphatic or heterocyclic amines.[4,35] Finally, as
mentioned above and emphasised here, detailed calculations
of the type described above have so far been justified
only with a conformationally rigid molecule.

References.

1. W.Naegele, Applications of High Field NMR Spectroscopy in Volume 4 of'Determination of Organic Structures by Physical Methods,'eds. Nachod and Zuckerman. Academic Press (1971).

2. D.R.Eaton and W.D.Phillips, Adv.Mag.Res.,$\underline{1}$, 103,(1965); E.de Boer and H.van Willigen, Prog.N.M.R.Spec., $\underline{2}$,111, 1967; H.J.Keller and K.E.Schwarzhans, Angew.Chem.,Int. Ed., $\underline{9}$,196, 946,(1970).

3. D.R.Eaton, J.Amer.Chem.Soc.,$\underline{87}$, 3097, (1965); E.R. Birnbaum and T.Moeller, ibid., 91, 7274, (1969).

4. F.A.Hart, J.E.Newbery, and D. Shaw, Chem.Comm., 45, (1967); J.Inorg. Nuclear Chem, $\underline{32}$, 3585, (1970).

5. C.C.Hinckley, J.Amer.Chem.Soc., 91, 5160, (1969).

6. J.K.M.Sanders and D.H.Williams. Chem.Comm., 422, (1970).

7. J.Briggs, G.H.Frost, F.A.Hart, G.P.Moss, and M.L. Staniforth, Chem.Comm., 749, (1970).

8. R.E.Rondeau and R.F.Sievers, J.Amer.Chem.Soc., $\underline{93}$, 1522, (1971).

9. G.M.Whitesides and D.W.Lewis, J.Amer.Chem.Soc., $\underline{92}$, 6279, (1970); $\underline{93}$, 5914, (1971).

10.K.G.Morallee, E.Nieboer, F.J.C.Rossotti, R.J.P.Williams A:V.Xavier, and R.A.Dwek, Chem.Comm., 1132,(1970).

11.F.A.Hart, G.P.Moss, and M.L.Staniforth, Tetrahedron Letters, 3389, (1971).

12.C.D.Barry, A.C.T.North, J.A.Glasel, R.J.P.Williams, and A.V.Xavier, Nature, $\underline{232}$, 236,(1971).

13. J.K.M.Sanders and D.H.Williams, Tetrahedron Letters, 2813, (1971).

14. Recent interviews include J.K.M.Sanders and D.H. Williams, Nature, 240, 385, (1972); M.R.Peterson jr., and G.H.Wahl jr., J.Chem.Educ., 49, 790, (1972); R. von Ammon and R.D.Fischer, **Angew**. Chem., Int.Ed., 11, 675, (1972).

15. D.R.Crump. J.K.M.Sanders, D.H.Williams, Tetrahedron Letters, 4419; (1970); N.Ahmad, N.S. Bhacca, J.Selbin, and J.D.Wander, J.Amer.Chem. Soc., 93,2564, (1971); W.De W.Horrocks jr. and J.P.Sipe III, ibid., p.6800.

16. C.S.Erasmus and J.C.A.Boeyens, Acta Cryst., B26, 1843, (1970).

17. J.P.R.de Villiers and J.C.A.Boeyens, Acta Cryst., B27 2335, (1971).

18. S.J.Schuchart Wasson, D.E.Sands and W.F.Wagner, Inorg.Chem., 12, 187, (1973).

19. J.J.Uebel and R.M.Wing, J.Amer.Chem.Soc., 94, 8910, (1972).

20. R.E.Cramer and K.Seff, Chem.Comm., 400, (1972).

21. J.S.Ghotra, F.A.Hart, G.P.Moss, and M.L.Staniforth, J.C.S.Chem.Comm., 113, (1973).

22. J.K.M.Sanders, S.W.Hanson, and D.H.Williams, J.Amer. Chem. Soc., 94,5325, (1972).

23. D.F.Evans and M.Wyatt, Chem. Comm., 312, (1973); N.H.Anderson, B.J.Bottino, and S.E.Smith, ibid., 1193; B.L.Shapiro and M.D.Johnson jr., J.Amer.Chem. Soc., 94, 8185 (1972).

24. B.Bleaney, J.Mag.Res., 8, 91. (1972).

25. J.Briggs, F.A.Hart, and G.P.Moss, Chem.Comm.,1506, (1970).

26. J.Briggs, F.A.Hart, G.P.Moss, and E.W.Randall, Chem. Comm., 364,(1971).

27. S.Farid, A.Atenza, and M.Maggio, Chem.Comm.,1285, (1971); M.R.Willcott tert., R.F.Leninski, and R.E. Davis, J.Amer.Chem.Soc., 94, 1742 (1972).

28. C.D.Barry, J.A.Glasel, A.C.T.North, R.J.P.Williams, and A.V.Xavier, Biochem.Biophys.Res. Comm., 47, 166, (1972).

29. J.M.Briggs, G.P.Moss, E.W.Randall, and K.D.Sales, J.C.S.Chem.Comm., 1180,(1972).

30. T.H.Siddall III, Chem.Comm., 452, (1971).

31. B.F.Pederson, Acta Chem.Scand., 21, 1415, (1967).

32. H.J.Schneider and W.Bremser, Tetrahedron Letters, 5197, (1970).

33. E.Wenkert, A.O.Clouse, D.W.Cochram, and D.Doddrell, J.Amer.Chem. Soc., 91, 6879,(1969).

34. see e.g. H.Hart and G.M.Love, Tetrahedron Letters, 625, (1971); I. Flemimg, S.W.Hanson, and J.K.M. Sanders, ibid., p. 3733.

35. R.J.Cushley, D.R.Anderson, and S.R.Lipsky, J.S.C. Chem.Comm., 636, (1972).

Table 1. Praseodymium induced shifts in the spectra of borneol.

Proton	2-exo	3-exo	3-endo	4	5-exo	5-endo	6-exo	6-endo	Me-8	Me-9	Me-10
Shift[a)	45.0	16.6	34.2	10.1	10.5	16.0	15.7	35.6	8.01	7.78	17.8
Calc.[b)	44.75	16.04	34.12	9.44	10.48	17.00	15.87	35,87	9.45	8.24	17.41

Carbon	C-1	C-2	C-3	C-4	C-5	C-6	C-7	C-8	C-9	C-10
Shift[a)	33.5	87.8	37.3	19.7	29.8	17.6	17.6	10.6	10.1	25.0
Calc.[c)	34.8	78.9	37.6	18.7	20.2	29.7	17.7	11.4	10.1	22.5

a) δ-value calculated for 1:1 mole ratio $Pr(tmhd)_3$: borneol minus δ-value for borneol alone.

b) Calculated as described in text. Pr-O = 3.00Å, Pr-O-C = 135.2°, Pr-O-C-C(1) = -179.8°.

c) Pr-O = 2.87Å, Pr-O-C = 126.1°, Pr-O-C-C(1) = -171.0°

Table 2. Europium induced shifts in the spectrum of (E) 2,6-diisopropyl acetanilide

Proton	CO–CH$_3$	Ar–CHMe$_2$	cis–CH$_3$	trans–CH$_3$
Shift[a]	9.4	4.1	1.74	-0.27
Calc.[b]	9.37	4.07 3.98	2.01 1.95	-.35 -0.16
		av. 4.03	av.1.98	av.0.26

a) see ref. 30   b) Calculated as described in text, Eu–O = 2.56 Å, Eu–O–C = 141°, Eu–O–C–Me = 177°.

Table 3. Praseodymium induced shifts for the lactone carboxylate (Fig.8).

(numbered as borneol see Fig.1.)

Proton	5 exo	5–endo	6–exo	6–endo	Me–8	Me–9	Me–10
Shift[a]	2.65	2.04	0.88	0.62	1.57	1.54	0.53
Calc.[b]	2.45	2.05	1.04	1.02	1.52	1.43	0.70

a) See Table 1 footnote a).   b) See text for details.

213

Table 4. Praseodymium induced shifts for the tricyclic ketone (Fig.9).

Methyl group	4α-Me	4β-Me	10-Me		
Observed shift[a]	8.86	11.36	5.06		
Conformation[b]					Pr-O
Calc.chair	10.44	7.76	3.50	2.97	Å
Calc.1,4α-boat	6.33	11.32	5.54	4.61	Å
Calc.1,4β-boat	8.54	10.52	2.77	4.97	Å
Calc.2,3,4,5-planar	7.63	10.95	3.15	2.97	Å

a) See Table 1 footnote a); only methyl signals given to illustrate
   results.

b) Separate calculations were conducted on the four conformations
   discussed in the text. The last one had Pr-O-C = 143°, Pr-O-C(4) = 172°.

Table 5. Selected lanthanide induced shifts with cedrol (See Fig.8)[a]

Carbon	C-8	2-Me	C-7	6S-Me	C-9	6R-Me
Eu shift	55.5	3.2	14.0	14.0	17.1	28.4
Pr shift	75.9	4.8	31.8	22.3	39.8	40.9
Ho shift	186.4	11.5	93.9	47.5	98.9	98.1

a) See Table 1 footnote a).

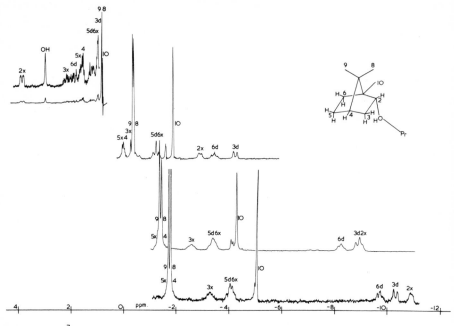

1. Top: ^{1}H n.m.r. spectrum of borneol in CCl_4 at 100 MHz.
   Below: the same but with increasing amounts of Pr(tmhd)$_3$

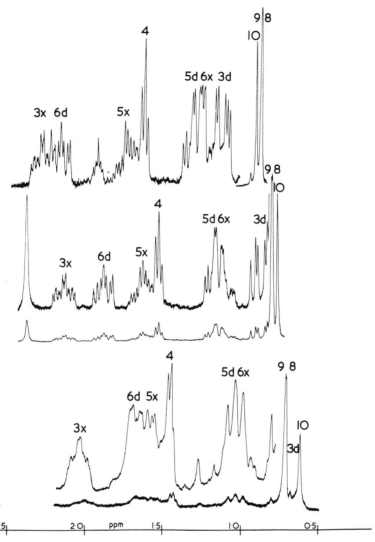

2. Middle: ^{1}H n.m.r spectrum of borneol in CCl_4 at 220 MHz.
Top: same with .018 mole proportion of $Eu(tmhd)_3$.
Bottom: same but with 0.012 mole proportion of
$Pr(tmhd)_3$.

217

3. Diagram of a 7-coordinate complex, $Ln(tmhd)_3 \cdot R_2CHOH$.

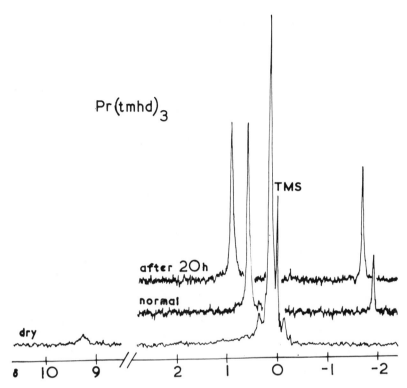

4. Bottom: [1]H n.m.r. spectrum of solution of freshly
   sublimed Pr(tmhd)$_3$.

Middle: spectrum of solution of sample previously stored
   as solid for several days in air.

Top: spectrum of solution after further storing in the
   air for 20 h.

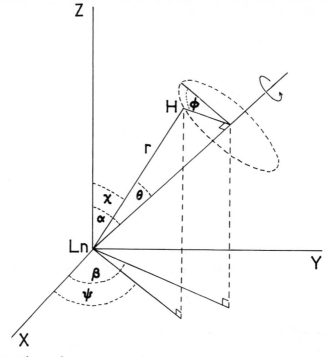

5. Relationship between susceptibility and rotation axes.

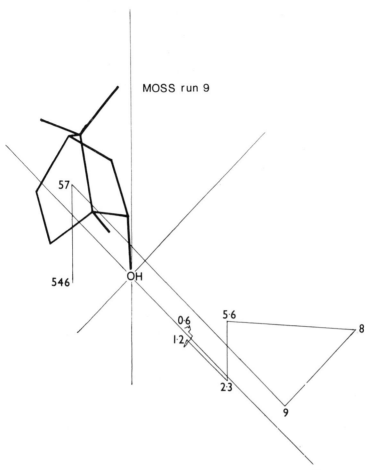

MOSS run 9

6. Graphic representation of computer program. The
numbers are R.M.S. derivations as defined in equation
(6).

7. Isomer E of 2,6-diisopropyl acetanilide.

8. The camphanate ion.

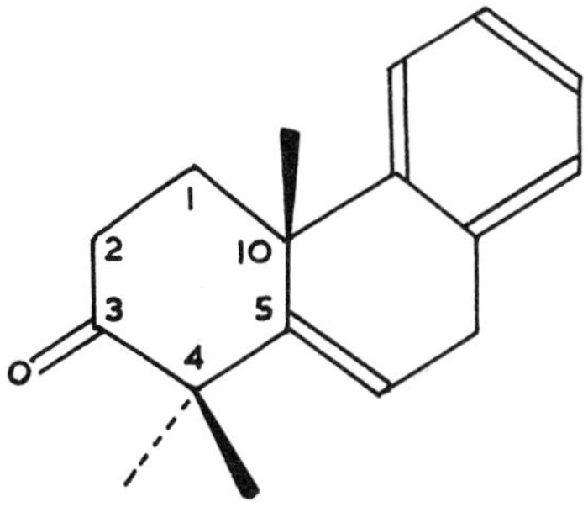

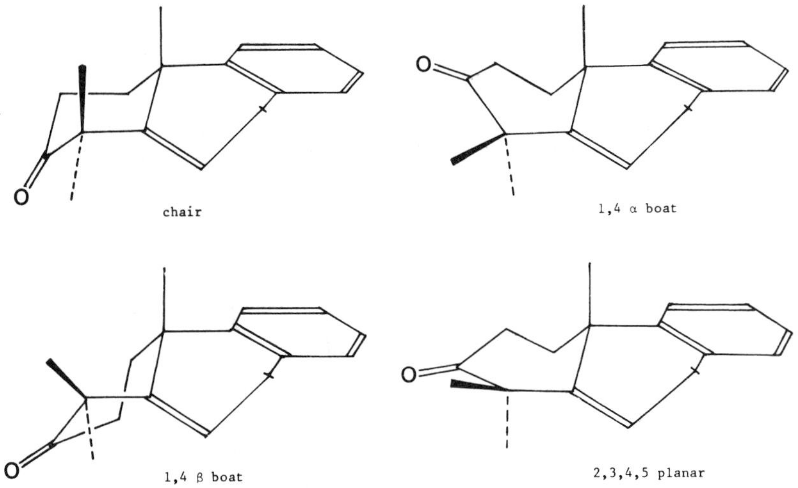

chair

1,4 α boat

1,4 β boat

2,3,4,5 planar

9. Structural formula and possible conformers of
podocarpa-5,8,11,13-tetra-3-one.

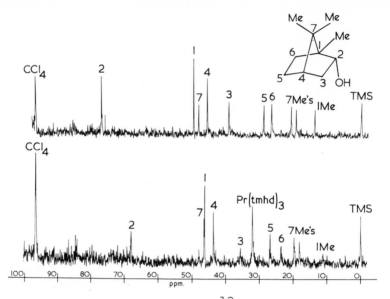

10.Top: proton noise decoupled ^{13}C spectrum of borneol.
Bottom: the same after addition of Pr(tmhd)$_3$ to soln.

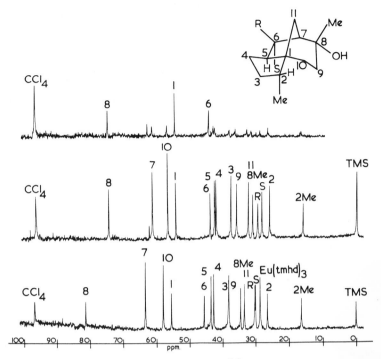

11. Middle: proton noise decoupled ^{13}C spectrum of cedrol.
    Top: "off-resonance" ^{13}C spectrum showing peaks from
        quaternary carbons.
    Bottom: spectrum obtained after addition of Eu(tmhd)$_3$
        to the solution.

Some Aspects of the Use of Lanthanide-Induced Shifts

In Organic Chemistry

B. L. Shapiro, M. D. Johnston, Jr.,[*] R. L. R. Towns,

A.D. Godwin,[†] H.L. Pearce,[§] T.W. Proulx[†] and M.J. Shapiro[†]

Department of Chemistry, Texas A&M University
College Station, Texas 77843

I. INTRODUCTION

Since their introduction by Hinckley less than four years ago, lanthanide shift reagents (LSR) and the lanthanide-induced shifts (LIS) these produce in NMR spectra have achieved an outstanding degree of acceptance by chemists working in a broad range of fields. Although much of the utility of LSR's derives simply from the dramatic increase in spectral dispersion they cause, most serious applications make use of the strong geographic dependence of the different magnitudes of the incremental shifts seen from a given molecule.

To rationalize these observed shifts, the usual procedure has been--in whole or in major part--to employ some form of the pseudocontact shift equation (1a), the simplest form of which is as follows:

$$\Delta\delta_i = k(3\cos^2\theta_i - 1)(R_i^{-3}) \qquad (1)$$

where $\Delta\delta_i$ is the incremental LIS of proton i, k represents

---

[*]Postdoctoral Fellow of The Robert A. Welch Foundation. Present address: Department of Chemistry, University of South Florida, Tampa, Florida.
[†]Predoctoral Fellow of The Robert A. Welch Foundation.
[§]Undergraduate Fellow of The Robert A. Welch Foundation.

a collection of constants, θ_i is the angle describing the position of proton i relative to the principal magnetic axis of the LSR, and R_i is taken as the proton-lanthanide ion distance.

According to equation 1 we obtain some measure of the incremental shifts $\Delta\delta_i$ and assay the assignments and/or molecular geometry using this simple two-parameter equation, with k being treated as a disposable scaling factor. (Various authors have also attempted to utilize modified, low-symmetry, forms of this equation as well as suggesting the possibility of introducing magnetic quadrupole and octupole moments (1b)). It must be emphasized that this approach has been *very* successful, and has led to the universal view that the pseudocontact contribution to the observed LIS is invariably highly predominant.

Considerable progress has also been made (2) in using the pseudocontact equation in concert with geometry-finding and -assaying computer programs widely used in x-ray crystallography, to make choices among various structural possibilities of molecules of known or assumed geometry.

These and other quantitative applications of LIS have recently encountered difficulties, for there have been several unpublished reports where pseudocontact shift equation modelling fails to reproduce precisely the known geometry of some molecules, with discrepancies of several percent being encountered. The practice which seems to be evolving is to blame these discrepancies on a relatively minor but still significant contribution to the observed shifts from a contact mechanism. It is, however, becoming quite clear that both pseudocontact and contact contributions may well be important where the LSR binding is to a hetero-atom directly involved in an extended pi-electron system, e.g., in pyridines (3). However, the imposition of such a contact contribution in order to repair geometry calculations for purely saturated systems and for protons (or carbons, etc.) several sigma bonds from the binding site seems to us to be on a substantially less firm ground theoretically and not as yet sufficiently documented experimentally.

Our own approach is one of attempting to make maximum use of the pseudocontact equation, or at least to be better

228

acquainted with the limitations of its applicability. We have chosen, for a variety of reasons, to work mostly with Eu(FOD)$_3$, despite the allegations that this lanthanide is more prone than others to inflict contact contributions on an otherwise simpler picture.

In looking at the pseudocontact shift equation we have chosen to focus our attention on *both* sides of equation 1. While the addition of a "tilt angle" or some additional parameters (to account for lower complex symmetries, different time-averaging requirements, etc.) may ultimately be needed on the right hand side of the equation, as well as a contact term, we are not at this time convinced of the general necessity for such measures. Clearly the conceptual and tactical attractiveness of the simple (equation 1) form of the pseudocontact shift mechanism is worthy of support and exploitation as long as possible.

Thus we have been concerned with determining the appropriate experimental measure(s) for the $\Delta\delta_i$'s for the left hand side of equation 1. Since such incremental shifts are always obtained from fast-exchange-limit-averaged spectra, we must examine the reaction stoichiometry with some care. Although the necessary reaction equilibrium theory is very ancient, its articulation vis-à-vis multistep equilibria has been, in our view, insufficiently exploited in NMR spectroscopic analysis. The consequences of this neglect have been, we feel, greater than hitherto expected.

We and others have previously shown ($\underline{4}$) that in at least some cases, a simple one-step binding process of the form L + S $\rightleftarrows$ LS is insufficient to explain the observed LSR concentration dependence of the shifts. Rather, a two-step mechanism of the form

$$L + S \rightleftarrows LS \qquad (K_1, \Delta_1) \qquad (2a)$$

$$LS + S \rightleftarrows LS_2 \qquad (K_2, \Delta_2) \qquad (2b)$$

is a *minimal* requirement. Once this is recognized, we are then faced with the problems and opportunities of determining (at least) *two* limiting incremental shifts and *two* equilibrium constants.

One must, then, make a decision about which incremental shift, Δ_1 or Δ_2, to use as the left hand side of the pseudo-contact shift equation in geometric assays. It seems to be no help to make arguments based on time-averaged equivalencing of the symmetry of the LS and LS_2 species, for we have found that in general the Δ_1/Δ_2 ratios are not constant for the various protons or carbons in a given compound. In any event, one must question the common practice of using raw initial slopes of (observed shift) vs. (LSR concentration) profiles as a uniform, adequate measure of the $\Delta\delta_i$'s needed for pseudocontact shift-based geometry calculations; as will be seen, this initial slope can in many cases of interest refer to Δ_1 or Δ_2 or to some blend of these. (It might also be mentioned that these initial slopes are often poorly determined and inadequately documented in published reports.) The resulting errors are of the same magnitude as, or even much larger than, those which have been blamed on contact contributions.

We have carried out reasonably careful LIS measurements by the "*constant-S_0, incremental dilution method*" (4a) and have analyzed the results by means of a new computer program LISA (Lanthanide-Induced Shifts Analysis) (5). Several types of results have been obtained, and we will proceed to discuss these along with our tentative conclusions. It must be emphasized in advance that these results and conclusions are extremely fresh, so that while we are fully confident of the major qualitative aspects, the details and extensions will remain under intensive investigation for some time.

II. EXPERIMENTAL METHODS

The LIS data are obtained by the *constant-S_0 incremental dilution method* (4a), in which the first experimental sample is prepared at the highest LSR concentration to be run, and subsequent samples are prepared by dilution with a substrate stock solution of the same substrate molarity as the first sample. This method provides very high precision in the concentrations and has been found to be satisfactory by us for reliable quantitative measurement of LIS. Usually twenty or more concentrations in the range $0 < \rho < 3$ (where $\rho = L_0/S_0$, the total molar concentrations of LSR and substrate, respectively) are run to assure good results. One of several simple tests for the quality of the data is a linear least-squares correlation coefficient of δ vs. ρ, at $\rho \lesssim 0.4$; a typical run will involve at least 7 or 8 such δ-values in this concentration range and these yield a linear regression coefficient, R, of at least 0.999 and frequently as good, or better than, 0.9999; a value of R smaller than 0.98 is cause for us to consider rejection of the data set. The simple application of this test also provides an automatic indication of the presence of any undesirable "scavenging" ($\underline{6}$). The details (and more general implications for other types of spectrometric analyses) of this incremental dilution technique will be published elsewhere in due course.

The resulting data points (ρ and δ_i for each proton or carbon) are then analyzed via LISA. Each resonance is fit separately to give the best values of the four parameters, by the method of least squares. Agreement of K_1 and K_2 values among the several protons of a typical molecule serves as a further check of the reliability of the data, as well as the precision and accuracy of the K's and Δ's so obtained.

The agreement between calculated shifts and observed data is gratifying; even for observed incremental proton shifts of up to 30 ppm, deviations between calculated and experimental values of the shifts are no larger than 0.1 ppm (0.33%) even in the worst cases. These fits are frequently much better than 0.1% (< 0.05 ppm) and the actual frequency values of the discrepancies are of course correspondingly much less for "slower moving" resonances. It must be emphasized, however, that obtaining such good agreement requires well-determined concentrations as well as chemical shifts of accuracies obtainable at sweep rates of 1 Hz/sec or less (at 100 MHz).

Other solution control strategies have been employed, of course ($\underline{7}$).

III. RESULTS AND DISCUSSION

The compounds we shall be concerned with here, with the exception of a few small substrate molecules such as cyclobutanone, cyclopentanone, and tetrahydrofuran, are highly substituted cyclohexanones ($\underline{1}$) and derived secondary

(2) and tertiary (3) methyl alcohols.

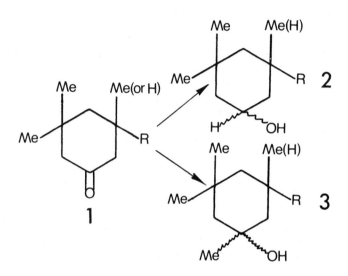

R = α- or β-naphthyl or various substituted phenyl groups

These unstrained, "normal" compounds afford a useful variety of definite stereochemistries (OH, R either axial or equatorial) and a considerable range of steric hindrance about the binding site. Some unanticipated conformational preferences have also been found (vide infra).

The Two-Step Equilibrium

There are now several independent lines of evidence (4) consistent with the common existence of two-step LSR binding cases (at least for Eu(FOD)$_3$), and these arguments will not be presented or evaluated here. It is sufficient for the present purposes to point out the superiority *in many cases* of the fit to a two-step scheme over that obtained from the best one-step fit; such a case is that of tetra-hydrofuran (with $Eu(FOD)_3$, CCl_4 solution) as illustrated in Figure I.

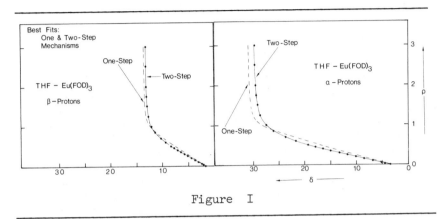

Figure I

In Fig. I the points are the actual observed shifts (the size of the points not reflecting the accuracy of the data). The two curves are the *best possible* (by least-squares criteria) one-step and two-step exact calculations. The initial slopes ($\sim 2\Delta_2$) of both the illustrated data sets show linear correlation coefficients of 0.99998 (for 10 points with $\rho \lesssim 0.4$); the obtention of excellent linear regions has been shown (4a) to be independent of the particular association stoichiometry which occurs. (It has been mentioned (4a) that self-association of the LSR may take place. The relatively low concentrations we employ minimize this effect so that our derived K's are off only by an activity coefficient close to 1. The Δ's are un-affected as are the ratios of the K's.)

Equilibrium Constants

In Table I are presented the results of measurements of the equilibrium (binding) constants between several sub-strates and $Eu(FOD)_3$ (in CCl_4 solution at $30°C$). As can be seen, the range of these is quite large, but reasonable in terms of the implied free energies. The standard deviation of typical results is $\pm$ 5% for K_1 and K_2, with most deter-minations being the average of the results for at least nine protons. The trends displayed by the compounds in

Table I are gratifying, but not surprising:

$$K_1(3^\circ \text{ alc.}) \; < \; K_1(\text{ketone}) \; << \; K_1(2^\circ \text{ alc.})$$

and

$$K_1(\text{axial OH}) \; < \; K_1(\text{equatorial OH})$$

Some further, qualitative, conclusions from the sizes of these K's should be pointed out here. The generally large values of the K's show that the LS species and at least some of the LS_2 complexes are quite stable, consistent with the many good existing geometry correlations. The non-parallel behavior between K_1 and K_2 (see the variations of K_1/K_2 in Table I) is consistent with the notion of a longer average L-S "bond" in the LS_2 complex than in LS. In LS_2, then, a looser - but still substrate geometry-sensitive - aggregate could mean that observed shifts may well depend on specific, detailed, degrees of internal freedom available to flexible parts of substrate molecules. Such differences, although having a relatively small impact on the free energy of complex formation, will function as a very sensitive probe of molecular geometry differences. We will amplify this point later.

Limiting Incremental Shifts

A selection of the many Δ_1 and Δ_2 values obtained from our compounds is given in Table II (Δ's are in ppm). These limiting-shift parameters, being intrinsic properties of the LS and LS_2 complexes, are the obvious candidates for potential use as *separate* $\Delta\delta_i$ shift measures in the pseudo-contact shift equation. They should be completely adequate for this purpose, for the accuracy of the limiting shift values is very high, in general as precise as the observed shifts themselves (± 2 Hz. at 100 MHz, at worst).

It is noteworthy that Δ_1 is always larger than Δ_2. This is clearly consistent with the arguments given above for the deduced relative sizes of K_1 and K_2, i.e., smaller shifts arising from the looser ternary complex, LS_2. A further reflection of this hypothesis is the finding of

234

	P	Q	V	U	C=O	Alcohol 3° ax	3° eq	2° ax	2° eq	K_1	K_2	K_1/K_2	$-\Delta G°(LS)$
1.	H	αN	Me				✓			48	2	21	2.3
2.	Me	αN	Me				✓			54	6	10	2.4
3.	αN	Me	Me				✓			81	1	57	2.6
4.	MA	Me	Me				✓			195	7	28	3.2
5.	PC	Me			✓					222	63	4	3.3
6.	βN	Me	Me				✓			271	3	93	3.4
7.	H	αN		Me		✓				612	65	9	3.9
8.	H	αN			✓					1,013	88	12	4.2
9.	βN	Me			✓					1,020	64	16	4.2
10.	αN'	Me		αN'		✓				1,097	150	7	4.2
11.	αN	Me			✓					1,896	90	21	4.5
12.	Me	αN	H						✓	87,000	9,600	9	6.8
13.	H	αN	H					✓		122,000	22,900	5	7.1
14.	H	αN	H						✓	218,000	24,500	9	7.4
15.	αN	Me	H						✓	228,000	21,500	11	7.4

αN = 1-naphthyl;   βN = 2-naphthyl;   MA = m-anisyl;   PC = p-chlorophenyl;
αN' = 1,2-naptho.   Values obtained in $Eu(FOD)_3/CCl_4$.

TABLE I.   EQUILIBRIUM CONSTANTS

non-uniform Δ_1/Δ_2 ratios for protons of the same substrate molecule (scan Table II horizontally). These departures from a constant proportionality of Δ_1 and Δ_2 clearly make the use of these different putative pseudocontact shift equation $\Delta\delta$'s non-equivalent. It is our present prejudice that Δ_1 values - *often not obtainable by simple inspection of initial slope values* (v.i.) - will prove to be the more useful for geometry calculations.

The data of Table II show a number of consistent trends, as well as providing the nucleus of a useful library of intrinsic, determination-independent LIS values. The large size and location-dependence of the observed Δ_1's should have useful predictive and correlative value. The internal consistency and the geographic type dependence of the shifts (again, better for Δ_1 than for Δ_2) is readily seen, for example, in comparing the axial methyl limiting shifts to those obtained for equatorial methyls in a common alcohol type, or shifts in alcohols compared to these ketones, etc. Similar correlations are in hand for the ring methylene group protons, and for a substantial number of aromatic hydrogens as well as several C-13 shifts.

Initial Slopes
<u>Initial Slopes</u>

Regardless of the particular solution preparation strategy employed (e.g., constant S_0, constant ρ, etc.), good quality data will yield a sizeable region in which the LIS are linearly dependent on the selected variable, for values of the latter near the undoped solution extreme (generally $\rho \stackrel{<}{\sim} 0.4$). Different control strategies result in linear regions of different extents and in different meanings of the resulting slopes. Under the conditions of our experiments (i.e., <u>at constant S_0</u>), the slope of the LIS vs. ρ plot is given to an high approximation by the following expression (<u>4a</u>):

$$\frac{\partial(\Delta\delta)}{\partial\rho} = 2\Delta_2 + \Delta_1/(S_0 K_2) \tag{3}$$

where $\rho < 0.4$ and $S_0 \stackrel{\sim}{\sim} 0.15M$ for most of the commonly encountered substrates in our work. The predominance of

TABLE II

Selected Values of Δ_1 and Δ_2 (ppm) for Highly Substituted Cyclohexanones and Cyclohexanols*

(Values Obtained in Eu(FOD)$_3$/CCl$_4$)

G→	Me$_x$ αN	βN	MA	Me$_z$ αN	βN	MA	Me$_y$ αN	βN	MA	Me$_v$ αN	βN	MA
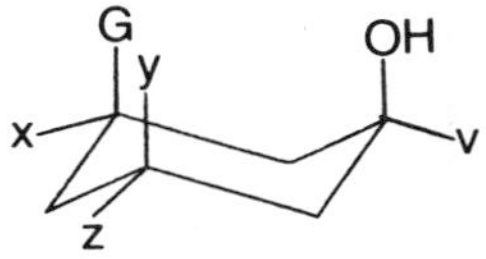 Δ_1	4.07	4.30	4.31	4.73	4.45	4.56	6.89	7.03	7.10	20.63	18.80	19.23
Δ_2	1.04	1.32	1.07	0.28	0.98	0.66	1.42	2.26	1.42	-0.40	4.56	3.35
Δ_1/Δ_2	3.9	3.3	4.0	17.	4.5	6.9	4.8	3.1	5.0	-51.	4.1	5.7
	αN			αN			αNy	αNy'		αN	βN	MA
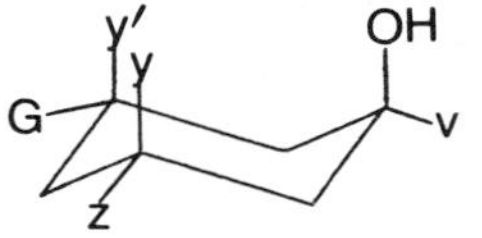 Δ_1				4.90			7.98	7.79			15.46	
Δ_2				2.09			3.44	3.54			6.87	
Δ_1/Δ_2				2.3			2.3	2.2			2.2	
	αN	βN	PCK	αN	βN	PCK	αN	βN	PC			
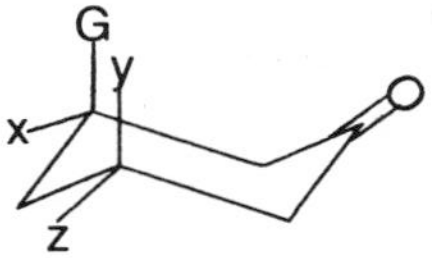 Δ_1	4.23	4.50	4.40	3.78	4.15	4.12	5.93	6.36	6.49			
Δ_2	2.03	1.81	1.84	1.81	1.62	1.70	3.02	2.88	2.90			
Δ_1/Δ_2	2.1	2.5	2.4	2.1	2.6	2.4	2.0	2.2	2.2			

*See bottom of Table I for the structure abbreviations.

237

TABLE II (Cont'd)

	H_g / Me_x / $Me_{y'}$	Me_y	Me_z	H_k
Δ_1	12.14	7.72	3.47	22.20
Δ_2	6.60	4.12	1.70	9.91
Δ_1/Δ_2	1.8	1.9	2.0	2.2
Δ_1	5.69	3.36	2.93	21.71
Δ_2	3.33	1.79	1.93	12.15
Δ_1/Δ_2	1.7	1.9	1.5	1.8
Δ_1	Me_x 3.18	4.07	3.01	21.39
Δ_2	1.91	2.17	1.76	12.86
Δ_1/Δ_2	1.7	1.9	1.7	1.7
Δ_1	$Me_{y'}$ 3.56	3.06	3.35	22.11
Δ_2	1.72	1.84	1.59	11.33
Δ_1/Δ_2	2.1	1.7	2.1	2.0

either the first or second term in the above equation can markedly simplify the interpretation of some sets of data and/or make it easier to obtain either (*but not both*) Δ_1 or Δ_2. A preliminary delineation of this application is seen in Table III.

	Restrictions on K_1 and K_2	Interpretation of Slope	Estimated Error in Slope	Some Compound Types
Type 1	a) $K_2 >$ ca. 50 or b) $(K_1/K_2) < 10$	$= 2\Delta_2$	ca. 5%	small and/or strongly bound substrates, e.g., THF (Table IV, No. 3), ketones (Table I, No. 5, Table IV, Nos. 1,2), 2^0 alcohols (Table I, Nos. 12-15).
Type 2	a) $K_2 <$ ca. 6 or b) $(K_1/K_2) > 50$ $K_2 <$ ca. 6	$\propto \Delta_1$	ca. 5 - 10%	weakly bound substrates, e.g. 3^0 alcohols (Table I, Nos. 1-3, 6).
Type 3	a) $K_2 = 7$ to 50? or b) (K_1/K_2) 10 to 50(?)	no high-approximation, simple, slope interpretation possible		intermediate strength binders, e.g. can include some hindered ketones and alcohols. See Table I.

TABLE III. INITIAL SLOPES.

Strong LSR Binding to Smaller Substrates: Increased Importance of LS_2

Another, different type of binding situation can be seen in the case of relatively small, essentially flat molecules which possess strong, sterically unhindered binding sites, e.g., ethers or ketones. In such cases one might expect larger K_2 values and smaller Δ_1/Δ_2 ratios than those seen for bulkier substrates. Examples of these expectations are found in Table IV, which presents data on three such substrates.

LIS and Conformational Analysis

For some time now we have held the view that LIS data can give valuable information about a wide variety of conformation problems. Several types of these results have been mentioned in our earlier publications in this area (6) and we wish to discuss here one novel type of such an LIS application.

In a recent paper dealing with compounds related to alcohol 3a, we presented evidence and arguments for the importance of conformation 3a relative to the seemingly

3a   3c   4b   4a

TABLE IV. Some Small Molecule Binding Data

N		K_1	K_2	K_1/K_2	Δ_1	Δ_2	Δ_1/Δ_2	$(\Delta_1/\Delta_2)^*$	σ	σ^*
CB-α	1.	540	50	10.2	13.26	7.44	1.78	1	8.9	6.7
-β					6.35	3.37	1.88	1.06	5.1	8.0
CP-α	2.	3400	200	17	15.50	9.00	1.72	1	3.4	2.2
-β					6.02	3.18	1.89	1.10	2.2	3.6
THF-α	3.	6248	180	34.7	26.36	16.49	1.60	1	3.8	1.4
-β					11.55	6.77	1.71	1.07	1.4	1.3

Values obtained in $Eu(FOD)_3/CCl_4$.

σ = standard deviation (in Hz at 100 MHz) between experimental points (22 for CB, CP; 24 for THF) and theoretical curve.

σ^* = above standard deviation normalized to a standard Δ_1 of 10 ppm.

CB = cyclobutanone.

CP = cyclopentanone.

THF = tetrahydrofuran.

energetically indistinguishable α-naphthyl orientations indicated by **3b** and **3c.** This evidence consisted in the observation that the observed LIS for protons 6 and 7 of the α-naphthyl system suffered *very substantial upfield* shifts, while all other protons suffered the usual downfield shifts associated with the more common angles which give rise to positive contributions from the pseudocontact equation angle factor. Similar shifts have also been found in the ^{13}C resonances of C-6 and C-7. We tentatively ascribe the preference for conformation **3a** to its markedly smaller molecular size relative to **3b** or **3c.** That the molecular energetics are sufficient to result in this conformational biasing seems subjectively reasonable, and we believe such findings to be quite novel.

Another such example is afforded by the axial β-naphthyl alcohol **4**, for which LIS data are given for all protons (except the OH) in Figure II. Here, the smaller conformation, **4a,** is clearly a more important contributor to the dynamic structure of this compound than is **4b.** Two specific pieces of evidence may be noted: (i) aromatic H-1 is seen to be shifted far more ($\Delta_1 = 676$ Hz) than is H-3 ($\Delta_1 = 402$ Hz). The substantial size of these shifts for this relatively weakly-binding, tertiary, axial alcohol clearly requires the dominance of conformations **4a** and **4b** (which are directly analogous to those previously demonstrated for an axial p-chlorophenyl group (8)); the large *upfield* LIS observed for H-4, H-5, H-6, H-7 and H-8 require the importance of conformation **4a** in order to provide the necessary negative angle factor in the pseudocontact shift dependence.

It is important to note that we have ruled out the very real worry that LSR-substrate binding might perturb the position of these rotational equilibria. This view is supported by spectral data obtained in the absence of LSR, as well as by noting that preferred conformations **3a** and **4a** are precisely those that would be suppressed if the LSR-substrate binding steric requirements perturbed the natural conformational equilibria of the substrates.

In both of these cases, then, the LIS data have been interpreted in terms of revealing the substantial involvement of molecular conformations differing primarily only

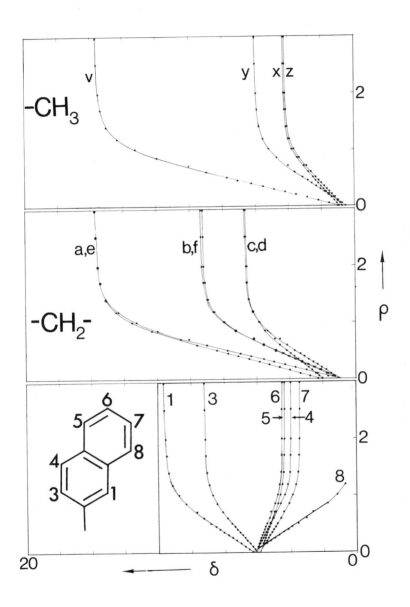

Figure II

in size, with the smaller conformer being favored. Such a finding is not only of considerable interest in itself, but indicates that LSR-substrate binding equilibria may well prove useful in examining a wide variety of low energy conformational preferences.

Acknowlegments

Major financial support for this work has been provided by The Robert A. Welch Foundation of Houston, Texas, for whose generosity we are deeply grateful.

References and Footnotes

1.(a) H. M. McConnell and R. E. Robertson, J. Chem. Phys., 29, 1361 (1958).

 (b) A. D. Buckingham and P. J. Stiles, Mol. Phys., 24, 99 (1972).

2.(a) M. R. Willcott, R. E. Lenkinski and R. E. Davis, J. Am. Chem. Soc., 94, 1742 (1972).

 (b) R. E. Davis and M. R. Willcott, ibid., 94, 1744 (1972).

 (c) J. Briggs, F. A. Hart and G. P. Moss, Chem. Commun., 1506 (1970).

 (d) C. D. Barry, A. C. T. North, J. A. Glasel, R. J. P. Williams and A. V. Xavier, Nature, 232, 236 (1971).

 (e) N. S. Angerman, S. S. Danyluk and T. A. Victor, J. Am. Chem. Soc., 94, 7137 (1972).

3. M. Hirayama, E. Edagawa and Y. Hanyu, Chem. Commun., 1343 (1972).

4.(a) B. L. Shapiro and M. D. Johnston, Jr., J. Am. Chem. Soc., 94, 8185 (1972).

(b) J. K. M. Sanders, S. W. Hanson and D. H. Williams, J. Am. Chem. Soc., 94, 5325 (1972).

(c) I. Armitage, G. Dunsmore, L. D. Hall and A. G. Marshall, Can. J. Chem., 50, 2119 (1972).

(d) J. W. ApSimon, H. Beierbeck and A. Fruchier, J. Am. Chem. Soc., 95, 939 (1973).

5.   Available on request.

6.   B. L. Shapiro, M. J. Shapiro, A. D. Godwin and M. D. Johnston, Jr., J. Mag. Res., 8, 402 (1972).

7.(a) D. R. Kelsey, J. Am. Chem. Soc., 94, 1764 (1972).

(b) T. A. Wittstruck, J. Am. Chem. Soc., 94, 5130 (1972).

(c) J. W. ApSimon and H. Beierbeck, Chem. Commun., 172 (1972).

8.   R. L. R. Towns and B. L. Shapiro, Cryst. Struct. Commun., 1, 151 (1972).

ORGANOMETALLIC ASPECTS OF
SHIFT REAGENT CHEMISTRY

T.J. Marks, R. Porter, J.S. Kristoff, and D.F. Shriver
Department of Chemistry, Northwestern University
Evanston, Illinois 60201

INTRODUCTION

Lanthanide β-diketonate shift reagents continue to have a great impact on the nmr spectroscopy of organic compounds.[1] It is the purpose of this chapter to summarize the considerable promise which shift reagents offer in the study of organometallic and coordination compounds.[2] Besides more conventional utilization for spectral simplification and stereochemical mapping, these reagents can also serve as unique magnetic resonance probes for functional group basicity.[2] For example, in a number of organometallic systems, nmr methods are considerably more convenient than experiments with boron halides, aluminum alkyls, etc. On a more general level, organometallic model compounds are useful in studying the complex equilibria necessary to describe shift reagent - substrate interactions in solution.[3]

CHEMICAL AND STRUCTURAL PROBES

Table I presents the results of adding $Eu(fod)_3$ (fod = 6,6,7,7,8,8,8 - heptafluoro - 2,2 - dimethyl - 3,5 - octanedionate) to solutions of a variety of organometallic and coordination complexes. Compounds containing halides and pseudohalides such as F, Cl, N_3 and CN were observed to interact with $Eu(fod)_3$ while those containing Br, I and NCS did not give observable shifts under the conditions of our experiments. For solutions containing $(\underline{h}^5-C_5H_5)Fe(CO)_2$ CN, the equilibrium mixture of coordinated and noncoordinated species was also observed in the infrared spectrum and the identity of the donor site, CN, was verified by the characteristic increase in $\nu(CN)$. The data for $(\underline{h}^5-C_5H_5)Mo(CO)_3Ge(X)(C_6H_5)_2$ where X = F, Cl indicate that fluoride coordinates preferentially to europium. These observations as well as others in this chapter demonstrate that the europium shift reagent is a hard Lewis acid, which is in harmony with general lanthanide chemistry.

The lack of observed shift for $(h^5-C_5H_5)Fe(CO)_2CH_3$ when contrasted with the observed shift for $[(h^5-C_5H_5)Fe(CO)_2]_2$ indicates that bridging metal carbonyl groups are more basic than terminal groups, in accord with previous results.[4] However, the dipolar shifts observed for (phen)-$[(C_6H_5)_3P]_2Mo(CO)_2$ are in accord with the notion that terminal carbonyls with sufficiently low CO stretching frequencies may serve as Lewis bases.[4] These results have provided the first evidence for a lanthanide interacting with a coordinated carbonyl. The basicity of $M-SO_2CH_3$ with H^+ and BF_3, and of $M-COCH_3$ complexes with H^+ is known,[5] however, the shift reagent experiments gave the first indication that a metal can interact with these coordinated ligands. Work in other laboratories has also demonstrated that standard basic functional groups on organic ligands are also susceptible to coordination by lanthanide shift reagents.[6]

Substantial shifts were observed with the transition metal base, $(h^5-C_5H_5)_2WH_2$, and with the main group metal base, $(h^5-C_5H_5)_2Sn$. In both cases, addition of tetrahydrofuran displaced the metal base from the lanthanide complex and regenerated the unshifted pmr spectrum of the metal base. These experiments have provided the first evidence for coordinate covalent transition metal-lanthanide and non-transition metal-lanthanide bonds.

The utility of the shift reagent for pmr spectral simplification in organometallic compounds is illustrated in Figure 1 which presents the result of adding $Eu(fod)_3$ to a toluene-d_8 solution of $(h^5-C_5H_5)Fe(COCH_3)(CO)P(C_4H_9)_3$. Dipolar shifts considerably simplify the pmr spectrum, while sharpness of the lines in the shifted spectrum permits detection of the spin-spin coupling between the phosphorus and cyclopentadienyl protons, proving that the iron-phosphorus bond has remained intact. Of interest also is the observation that only certain multiplets in the phenyl pmr resonances of phenylarsine and phosphine (cis to halogen) complexes are shifted upon addition of $Eu(fod)_3$. The spatial dependence of the dipolar shift implies that these resonances are due to ortho protons on the phenyl rings.

A serious limitation of all the β-diketonate lanthanide shift reagents is their lack of interaction with soft bases, e.g. phosphines, sulfides[1,7]. This presumably reflects the hardness of lanthanide ions in the +3 oxidation state, amplified by the hardness of the anionic oxygen-containing ligands. In an effort to circumvent this limitation, lanthanide complexes with presumably softer, π-bonding ligands were tried as shift reagents. As can be seen in Figure 2, the complex $(\underline{h}^5\text{-}C_5H_5)_3Yb$[8] induces large isotropic pmr shifts in spectra of soft bases such as $(n\text{-}C_4H_9)_3P$ and $(C_2H_5)_2S$. Under the same conditions, the corresponding fod complex produced shifts which were smaller by at least an order of magnitude. Drawbacks of these organometallic shift reagents include increased air and moisture sensitivity as well as reduced solubility compared to the fod analogs; also spectral broadening is greater at all substrate:shift reagent ratios. Further work will be necessary to determine to what extent these effects are due to hardness or softness as contrasted with intrinsic acidity and differences in magnetic anistropy. However, these results suggest that it may be possible to synthesize "tailor-made" shift reagents by rational manipulation of ligand bonding characteristics.

STOICHIOMETRY PROBES

Central to the quantitative application of nmr shift reagents as structural and chemical probes are the stoichiometry and equilibria of interaction between lanthanide shift reagent, Ln, and substrate, S. Surprisingly little is actually known about the identity of the species which exist at ambient temperature in typical reagent-substrate solutions. Until the recent experiments of Shapiro and Johnston[9], analyses of substrate chemical shifts as a function of lanthanide:substrate ratio have usually[10] (though not always[11]) supported 1:1 adduct formation at room temperature, and in some cases, equilibrium constants have been calculated. Shapiro and Johnston's nmr analysis of $Eu(fod)_3$ - substrate equilibria yielded successive formation constants for both 1:1 and 1:2 adduct formation. The inherent possible complexity of the equilibria occurring in these systems indicated to us that an alternative physiochemical technique, in this case vapor pressure osmometry, would be a useful if not necessary supplement to even the most rigorous analysis of substrate nmr shift data. The possibility of extensive shift reagent association or 1:3 (or even 1:4) adduct formation cannot, a priori, be ruled out in shift reagent-substrate solutions.

We present here correlated osmometric and pmr studies with the model substrate $(\underline{h}^5-C_5H_5)Fe(CO)_2CN$, which was chosen because it displays a single sharp resonance in the nmr, the site of coordination is known,[2] and it can be accurately weighed. Because of recent interest in using lanthanides other than europium for shift reagents, we present data on the systems $Ln(fod)_3$, where $Ln = Pr$, Eu, Ho, and Yb. The scope of equilibria occurring is considerably richer than heretofore believed.

Figures 3 and 4 present representative chemical shift data as a function of the ratio, Ln/S, where Ln denotes shift reagent concentration and S, substrate concentration. These curves are similar to numerous ones in the literature, where the inflection in the curve at $Ln/S \approx 1$ was ascribed to the formation of 1:1 adducts. Figure 3 also illustrates the effect of neglecting to remove moisture and other impurities from $Eu(fod)_3$. Clearly, shifts are reduced and the curve does not break as sharply. Tables II and III present the results of refining osmometric molecular weight data on solutions similar to those employed for the nmr measurements. The following sets of mass balance equations were employed to analyze the osmometric data in a quantitative fashion. The subscript tot denotes the total quantity in solution. M_{tot} denotes total solute. For shift reagents without substrate,

$$M_{tot} = Ln + (Ln)_2$$

$$Ln_{tot} = Ln + 2(Ln)_2$$

$$K_D = \frac{\left[(Ln)_2\right]}{\left[(Ln)\right]^2}$$

Table II indicates that the association of the pure shift reagents in benzene falls off considerably with increasing atomic number of Ln. Since the first equilibrium constant for substrate - shift reagent interaction was immeasurably large ($K_{1:1} > 10^3$), data could only be analyzed in the region where $Ln/S < 1$,

$$S_{tot} = S + LnS + 2LnS_2$$

$$Ln_{tot} = LnS + LnS_2$$

$$M_{tot} = S + LnS + LnS_2$$

$$K_{1:1} = \frac{[LnS]}{[Ln][S]} > 10^3$$

$$K_{1:2} = \frac{[LnS_2]}{[LnS][S]}$$

Equilibrium constant data, for the Eu, Ho, and Yb systems along with estimated errors, are given in Table III. Both 1:1 and 1:2 complexes are present. As can be seen in the Table, $Pr(fod)_3$ displays even greater propensity for association. Indeed, both $K_{1:1}$ and $K_{1:2}$ were immeasurably large. However, $K_{1:3}$ could be determined in the region $Ln/S < 0.5$, using

$$S_{tot} = S + 2 LnS_2 + 3 LnS_3$$

$$Ln_{tot} = LnS_2 + LnS_3$$

$$M_{tot} = S + LnS_2 + LnS_3$$

$$K_{1:3} = \frac{[LnS_3]}{[LnS_2][S]}$$

251

For the Eu, Ho, and Yb systems, it is also possible to
calculate the resonance position (δ) of S in LnS and in
LnS$_2$ by combining the osmometric results with the pmr
data. The result is presented in Table III. Invariably,
the C$_5$H$_5$ protons are shifted to a greater extent in the
1:1 adducts. Such stoichiometric determinations are by no
means restricted to $(\underline{h}^5\text{-}C_5H_5)Fe(CO)_2CN$ as the substrate.
The last entry in Table III is for solutions of Eu(fod)$_3$
and cholesterol, which demonstrates the presence of 1:2
Ln:S adducts in this particular system.

The shape of the curve for nmr shift <u>versus</u> Ln/S ratio
has been the source of much confusion with respect to the
stoichiometry of the lanthanide-substrate interaction.
Superficially, a typical curve such as that presented in
Figure 3 and 4 might be interpreted in terms of simple 1:1
interaction of Ln and S. However, the osmometry results
clearly reveal the formation of 1:2 and in some cases
higher complexes. There are several reasons for the relatively simple shape of the nmr shift curves. Most important is the fact that the measured shift is an average
of shifts for the various substrate-containing species,
S, LnS, LnS$_2$, LnS$_3$, etc.. In addition, the combination of
osmometry and nmr data shows that for the present case the
successive formation constants decrease markedly from
$K_{1:1}$ to $K_{1:2}$, and that the chemical shifts for the 1:1
adduct are equal to or somewhat greater than the shift for
the 1:2 adduct. As a result, the progression from the low
Ln/S ratio represents (for Eu, Ho and Yb) initially the
formation of LnS$_2$, which upon addition of Ln is converted
to LnS with a smooth increase in the observed average
chemical shift. When $K_{1:1}$ is large and $K_{1:2}$ is significantly less, the curve eventually will level off around
Ln/S of 1.0 because around this point all of the substrate
is present as 1:1 and no further change occurs upon addition of shift reagent. (We are neglecting the formation
of Ln$_2$S, for which there appears to be some evidence in
the case of Pr.) When $K_{1:2}$ is fairly large the break will
occur before the 1:1 point, as for example, with Eu(fod)$_3$,

(Figure 3). Our results appear to be in reasonable agreement with the most detailed nmr analysis to date[9] (which employed a different substrate). The rather featureless character of shift $\underline{vs}$. Ln/S curves combined with the large number of unknowns, requires extremely accurate nmr data and a rather formidable numerical analysis to obtain information for successive 1:1 and 1:2 complex formation. The strength of osmometry lies in the possibility of recognizing the formation of higher adducts (e.g., LnS_3) in addition to 1:1 and 1:2 species. The determination of very large equilibrium constants is beyond the capability of the osmometry method.

The osmometry data also reveal some interesting periodic aspects of lanthanide tris(β-diketonate) coordination chemistry. It is apparent from the available structural data that steric interactions are important factors in the structural chemistry of shift reagent systems. Our equilibrium results are in close accord with changes in ionic radius [12] across the lanthanide series, namely that the heavier metals less readily undergo expansion of the coordination sphere.

The effectiveness of a given shift reagent system is a complex function of magnetic anisotropy, electron spin relaxation time, solubility, and substrate affinity. Our results indicate that the latter characteristic cannot always be ascertained accurately by nmr methods alone, and that complementary colligative studies may be helpful in guiding systematic efforts to design more stereoselective shift reagents.

Acknowledgements: This research was supported by the NSF through grants GP-28878 (D.F.S.) and GP-30623X (T.J.M.) R. Porter is the recipient of an NDEA fellowship.

References

1. R. von Ammon and R.D. Fischer, $\underline{Angew}$. $\underline{Chem}$. $\underline{Int}$. $\underline{Ed}$. $\underline{Engl}$., $\underline{11}$, 675(1972); W. DeW. Horrocks, $\underline{Jr}$., and $\underline{J.P}$. Sipe, III, $\underline{J}$. $\underline{Amer}$. $\underline{Chem}$. $\underline{Soc}$., $\underline{93}$, 6800(1971).

2. T.J. Marks, J.S. Kristoff, A. Alich, and D.F. Shriver, $\underline{J}$. $\underline{Organometal}$, $\underline{Chem}$., $\underline{33}$, C35(1971).

3. R. Porter, T.J. Marks, and D.F. Shriver, $\underline{J}$. $\underline{Amer}$.

Chem. Soc., in press.

4. D.F. Shriver, Chemistry in Britain, 8, 419(1972).

5. R.A. Ross and A. Wojcicki, Inorg. Chim. Acta, 5, 6(1971); M.L.H. Green, L.C. Mitchard, and M.G. Swanwick, J. Chem. Soc., A, 794(1971).

6. M.J. Foreman and D.G. Leppard, J. Organometal. Chem., 31, C31(1971).

7. J.K.M. Sanders and D.H. Williams, Tet. Letters, 2813 (1971); T.J. Marks, R. Porter, and D.F. Shriver, unpublished results.

8. F. Calderazzo, R. Pappalardo, and S. Losi, J. Inorg. Nucl. Chem., 28, 987(1966); E.O. Fischer and H. Fischer, J. Organometal. Chem., 6, 141(1966).

9. B.L. Shapiro and M.D. Johnston, J. Amer. Chem. Soc., 94, 8185(1972).

10. J. Armitage, G. Dunsmore, L.D. Hall, and A.G. Marshall, Chem. Comm. 1281(1971).

11. J.W. ApSimon, H. Beierbeck, and A. Fruchier, J. Amer. Chem. Soc., 95, 939(1973).

12. T. Moeller in "MTP International Review of Science, Inorganic Chemistry", Ser. 1, Vol. 7, H.J. Emeleus and K.W. Bagnal, Eds., University Park Press, Baltimore, Md., 1972, p. 275; G.R. Chopin, Pure Appl. Chem., 27, 23(1971).

TABLE I

Observed Isotropic Shifts for Metal Complexes with $Eu(fod)_3$ [a]

Compound	Concen.	Concen. $Eu(fod)_3$	Max Recorded[b] Shift of Protons	
$(h^5\text{-}C_5H_5)_2Ti(N_3)_2$	0.31	0.30 M	C_5H_5	0.47 ppm
$[(C_6H_5)_3P]_2Ir(CO)Cl$	0.10	0.30	ortho	0.10
			meta, para	0
$(h^5\text{-}C_5H_5)_2TiCl_2$	0.16	0.34	C_5H_5	0.10
$[(CH_3)_2(C_6H_5)As]_2Ru(NO)Cl_3$	0.26	0.27	CH_3	0.17
			ortho meta,	0.07
			para	0
$(h^5\text{-}C_5H_5)Fe(CO)(COCH_3)P(n\text{-}C_4H_9)_3$	0.33	0.35	C_5H_5	3.62
			$COCH_3$	3.56
			$\alpha\text{-}CH_2$	3.82
			$\beta\text{-}CH_2$	2.60

255

TABLE I (cont.)

			Y	
			γ-CH$_2$	0.50
			CH$_3$	0.27
$(\underline{h}^5-C_5H_5)Mo(CO)_3Ge(C_6H_5)_2F$	0.33	0.17	C$_5$H$_5$	0.55
			ortho	1.03
			$\underline{meta},\ \underline{para}$	0.07
$(\underline{h}^5-C_5H_5)Mo(CO)_3Ge(C_6H_5)_2Cl$	0.16	0.30	C$_5$H$_5$	0
			ortho	0.03
			$\underline{meta},\ \underline{para}$	0.01
$(\underline{h}^5-C_5H_5)Fe(CO)_2CH_3$	0.33	0.23	CH$_3$	0
			C$_5$H$_5$	0
$(\underline{h}^5-C_5H_5)Fe(CO)_2CN$	0.10	0.22	C$_5$H$_5$	3.81
$(\underline{h}^5-C_5H_5)Fe(CO)_2Cl$	0.10	0.34	C$_5$H$_5$	3.12
$(\underline{h}^5-C_5H_5)Fe(CO)_2Br$	0.10	0.19	C$_5$H$_5$	0

TABLE I (cont.)

$(h^5-C_5H_5)Fe(CO)_2I$	0.10	0.15	C_5H_5	0
$(h^5-C_5H_5)Fe(CO)_2NCS$	0.10	0.07	C_5H_5	0
$(h^5-C_5H_5)Fe(CO)_2(SO_2CH_3)$	0.10	0.01	CH_3	0.87
			C_5H_5	0.53
$\left[(h^5-C_5H_5)Fe(CO)_2\right]_2$	0.032	0.12	C_5H_5	0.35
$(phen)\left[P(C_6H_5)_3\right]_2Mo(CO)_2^c$	0.018	0.055	broad feature	0.27
			broad feature	0.48
$(dmp)\left[P(C_6H_5)_3\right]_2Mo(CO)_2^c$	0.023	0.030		ca. 0.18
$(h^5-C_5H_5)_2Sn$	0.65	0.17		0.75

257

TABLE I (cont.)

| $(\underline{h}^5\text{-}C_5H_5)_2WH_2$ | 0.19 | 0.14 | C_5H_5 | 1.78 |
| | | | H | -2.11 |

[a] Data were obtained at 60 MHz for complexes dissolved in toluene-d_8 or chloroform-d_1 with TMS as an internal reference. Maximum recorded shifts (not necessarily the limit at very high Eu(fod)$_3$ concentrations) are presented.

[b] Positive entries represent downfield shifts. The estimated precision is $\pm$ 0.025 ppm.

[c] Decomposes fairly rapidly in CDCl$_3$ solution. phen = 1,10-phenanthroline; dmp = 5,6-dimethy-1,10-phenanthroline.

TABLE II

OSMOMETRIC RESULTS FOR DIMERIZATION OF

$Ln(fod)_3$ SHIFT REAGENTS AT 37° IN BENZENE

Metal	K_D [a]	% Dimerized[b]
Pr[c]		
Eu[d]	60.9 $\pm$ 18	33.4
Dy	13.8 $\pm$ 4.0	21.9
Ho	11.6 $\pm$ 3.8	20.4
Yb	8.6 $\pm$ 3.6	17.8

[a] In M^{-1}

[b] At total $Ln(fod)_3$ = 0.05 M

[c] Concentration dependence of K_D indicates aggregates greater than dimers are also present.

[d] Shows slight concentration dependence at concentrations greater than 0.07 M.

Table III

OSMOMETRIC RESULTS FOR SHIFT REAGENT AND SUBSTRATE AT 37° IN BENZENE

Metal	Substrate	$K_{1:1}$ [a]	$K_{1:2}$ [a]	$K_{1:3}$ [a]	$\delta_{1:1}$	$\delta_{1:2}$
Pr	$CpFe(CO)_2CN$ [d]	$>10^3$	10^3	25.7 ± 13		
Eu	$CpFe(CO_2)CN$	$>10^3$	290 ± 145		8.2 [b]	6.9 [b]
Ho	$CpFe(CO)_2CN$	$>10^3$	80.7 ± 40		34.1 [c]	18.9 [c]
Yb	$CpFe(CO)_2CN$	$>10^3$	39.2 ± 20		18.1 [b]	12.5 [b]
Eu	Cholesterol	$>10^3$	93.8 ± 45			

[a] In M^{-1}

[b] In ppm downfield from TMS

[c] In ppm upfield from C_6H_6

[d] $Cp = h^5-C_5H_5$

260

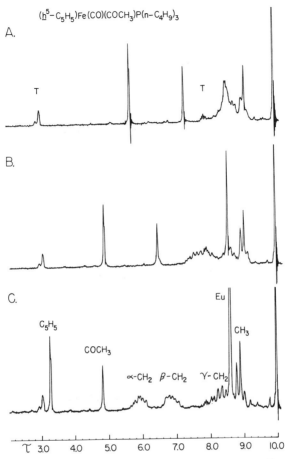

Fig. 1. A. Proton nmr spectrum (60 MHz) of $(\underline{h}^5\text{-}C_5H_5)Fe$
(CO)(COCH$_3$)P(n-C$_4$H$_9$)$_3$ as a 0.33 M solution in
toluene-d$_8$. Peaks marked T are due to traces
of toluene-d$_7$.

B. Sample A, containing 0.077 M Eu(fod)$_3$.

C. Sample A, containing 0.22 M Eu(fod)$_3$.
Spectral assignments are as shown. The peak
marked Eu is due to the methyl groups of the
shift reagent.

261

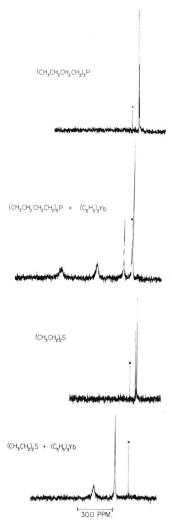

Fig. 2. Proton nmr spectra (60 MHz) illustrating the effect of adding a <u>ca</u>. 25% molar excess of $(C_5H_5)_3Yb$ to benzene-d_6 solutions of the indicated bases. The "X" denotes the resonance of benzene-d_5.

262

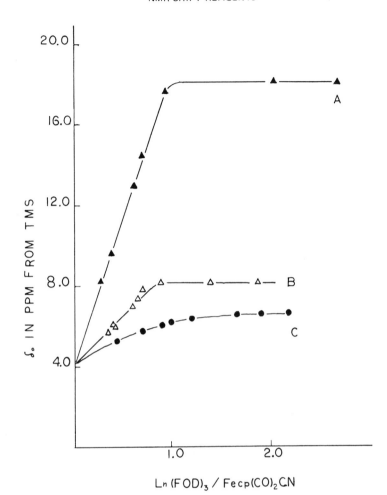

Fig. 3. Observed downfield shift of $(C_5H_5)Fe(CO)_2CN$ from TMS plotted against shift reagent to substrate ratio at 37°C. Concentrations are the same as for osmometric studies (Ln = 0.040M). A. Ln = Yb; B. Ln = Eu with exclusion of H_2O; C. Ln = Eu without exclusion of H_2O.

263

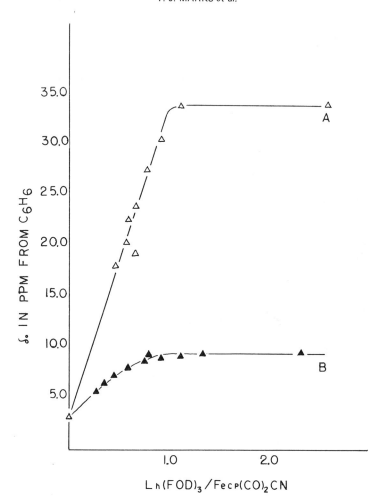

Fig. 4. Observed upfield shift of $(C_5H_5)Fe(CO)_2CN$ from
benzene plotted against shift reagent to substrate
ratio at $37^{\circ}C$. Concentrations are the same as
for osmometric studies (Ln = 0.040M). A. Ln =
Ho; B. Ln = Pr.

CHEMICALLY INDUCED DYNAMIC NUCLEAR POLARIZATION IN THE PRESENCE OF PARAMAGNETIC SHIFT REAGENTS.

Joachim Bargon

IBM Research Laboratory, Molecular Physics Department,
San Jose, California 95193

Abstract

Paramagnetic shift reagents can be used to selectively shift the NMR lines of reaction products showing the CIDNP phenomenon. If small amounts of the reagents are added, the CIDNP intensities are not altered significantly, whereas large amounts quench the phenomenon. Overlapping lines can be shifted apart. This technique is demonstrated during the decomposition of benzoyl propionyl peroxide in CCl_4 in presence of $Eu(fod)_3$ and $Pr(fod)_3$. Reaction products which destroy these shift reagents have to be avoided. Because of changing molar concentration ratios between the lanthanide complexes and the products during the reactions, the line positions move as a function of the reaction time. In photochemically initiated reactions the shift reagents act as very efficient quenchers and are of very limited value.

Despite the success of shift reagents in absorption – NMR spectroscopy it is not straightforward to extrapolate their performance to systems generated with nuclear-spin polarization, i.e., to those which show emission and enhanced absorption lines in the NMR spectra. A technique that yields NMR emission lines is, for example, the Chemically Induced Dynamic Nuclear Polarization (CIDNP) method. The CIDNP phenomenon was discovered accidentally during free-radical polymerization reactions[1], and later investigated in more detail during the thermal decomposition of peroxides[2]. It was also indepently observed during reactions of alkyl halides with alkyl lithium compounds[3]. It was immediately realized that the CIDNP phenomenon was correlated with the occurrence of free radicals, and it has been shown that the products resulting from free-radical intermediates are initially formed with a significant nuclear-spin polarization which causes the intense emission lines[3].

265

A typical example is shown in figure 1, recorded during
the thermal decomposition of dibenzoyl peroxide (BPO) in
cyclohexanone as the solvent[2]. At first (t=0) the normal
absorption spectrum of the BPO is observed. As the sample
warms up an emission line is observed (t=4 min) at the
characteristic chemical shift of benzene, which is known
to be a product of this reaction. The intensity of the
benzene-emission line reaches a maximum, decreases, vanishes after about 8 minutes and reappears later (t=12 min)
as an absorption line. Figure 2 shows the change from emission to absorption and the intensities of the benzene resonance during the reaction.

The presence of paramagnetic species other than free-radical intermediates during CIDNP-showing reactions has been
noticed to cause a drastic decrease, for example, of the
benzene emission intensity during the above reaction[4]. Figure 3 shows the influence of Fe^{3+} ions on the intensity
plot of the benzene resonance during the BPO decomposition.
Accordingly, the presence of any paramagnetic species reduces the CIDNP intensities, and a similar consequence can
be expected from paramagnetic shift reagents.

The Origin of CIDNP

The nuclear-spin polarization found in the reaction products has been shown to originate from radical pairs [5],
which are the precursors of the products. If the pairs are
formed by statistical encounters of two free radicals,
they can either have an electronic singlet or triplet
character. Of these two possibilities only the singlet
pairing can yield products. Triplet pairs have to separate
again or to undergo intersystem crossing to the singlet
state. Such an intersystem crossing in a radical pair can
be visualized in a simple classical picture as a dephasing
process: In the strong magnetic field of the NMR spectrometer the unpaired electrons of the radical pairs precess
around the direction of the field with their Lamor or ESR
frequency ω. If the two radicals in a pair are different,
i.e., when they have different g-values, they will precess
with different frequencies ω_1 and ω_2. The consequences of
a precession - frequency difference $\Delta\omega = \omega_2 - \omega_1$ in a radical pair, in which the unpaired electrons are aligned
antiparallel to one another, is illustrated in figure 4.

266

If the radical pair is initially in its singlet state (S_O), then the two electrons will precess 180 ^{O} out of phase (left). Because of $\Delta\omega$ the electrons will lose their phase correlation and will eventually reach the situation, where they are precessing in phase (right). This phase correlation, however, corresponds to a triplet state (T_O) of the pair. The rate of the dephasing, which is an oscillating process, depends on the size of $\Delta\omega$. Its value is determined among other things by the orientation of the magnetic nuclei in the radicals. Their magnetic moments cause a local magnetic field, which can either add to or subtract from the external magnetic field, thus defining an effective field, which can be different for the two unpaired elec-- trons. Consequently the two electrons will precess with individual precession frequencies, which can be higher or lower than that of the free electron.

The effective magnetic field is determined by three parameters: 1.) the amount of spin-orbit coupling in the radical, i.e., by its g-value, 2.) by the size and the sign of the hyperfine-coupling constant a_H, and 3.) by the magnetic quantum number of the nuclei in the radicals, which indicates their orientation. The precession-frequency difference reaches a maximum, if the magnetic nuclei, for example the protons, having a positive a_H are in the $m_I = - 1/2$ (energetically higher) state in the radical with the higher g-value but in the $m_I = + 1/2$ (energetically lower) state when in the radical with the lower g-value. A consequence of these relations is that, for example, triplet radical pairs having more favorable oriented protons will dephase faster and thus be able to form products through combining earlier than the others with less favorably oriented protons. This results in a nuclear-spin sorting in radical pairs, which is subsequently transferred to the products as a nuclear-spin polarization, where it can be observed in form of emission or enhanced absorption lines in the high-resolution NMR spectra.

As singlet-radical pairs can combine or disproportionate to (singlet) products, whereas the triplet pairs cannot but may separate and attack the solvent to form the (triplet) transfer products, for example by addition or abstraction reactions, the singlet and the triplet products show a complementary nuclear spin polarization as a direct conse-

quence of the nuclear-spin sorting in the intersystem crossing step. Thus if singlet products show emission, the same
nuclei in the triplet products will show enhanced absorption, and vice versa. Whatever is to be expected can be predicted from the above parameters by means of a simple product rule[6], which connects the options of four parameters
like the multiplication rules for plus and minus signs.

$$M \cdot a_H \cdot \Delta g \cdot P = \left\{ \begin{array}{l} + \rightarrow \text{Absorption} \\ - \rightarrow \text{Emission} \end{array} \right.$$

In this rule M is the multiplicity of the radical pair in
its initial stage with the + sign for triplet pairs and the
– sign for singlet pairs. a_H is the sign of the hyperfine –
coupling constant of the proton under investigation. Δg
stands for the g–value difference in the radical pair; a +
sign applies for the protons in the radical with the higher
g–value, a minus sign for those in the other radical. P is
the type of the product. For singlet products, i.e., for the
combination and disproportionation products, the + sign is
taken, whereas the – sign stands for triplet products. The
sign of the product symbolizes emission or enhanced absorption.

Figure 5 illustrates the origin of CIDNP schematically in
form of an energy diagram for protons with positive a_H in
a radical with the lower g–value in a pair of singlet multiplicity. The set of parameters is accordingly M = –,
a_H = +, Δg = –, and the result is + (absorption) for combination and disproportionation products (P = +) but – (emission) for transfer products (P = –). In an analogical way
the CIDNP expected in reactions involving any kind of radical pair can be predicted with the above formula as long as
there is a g–value difference in the pair, and if the reactions are run in high magnetic fields.

CIDNP and Shift Reagents

Since it is known that processes which involve a change in
spin multiplicity proceed appreciably faster in the presence of paramagnetic materials[7], it is an interesting question whether shift reagents interfere with the intersystem
crossing in free-radical pairs, which is the very source of
CIDNP. Therefore, we have investigated the thermal decomposition of the asymmetric benzoyl propionyl peroxide (BPPO),

268

$$C_6H_5 - CO - O - O - CO - CH_2 - CH_3$$

which is otherwise similar to the symmetric peroxide BPO discussed before, in the presence of shift reagents in CCl_4 as the solvent. The details of its decomposition mechanism and the CIDNP during this reaction has been discussed elsewhere[8]. Figure 6 shows that part of the reaction mechanism that leads to the products ethyl benzoate, ethyl chloride, and ethylene, the CH_2 resonances of which are shown in figure 7 recorded during the reaction. At 75° C BPPO decomposes under CO_2 loss into a pair of a benzoyloxy and an ethyl radical. Because the pair results from a diamagnetic precursor, it is initially formed with electronic singlet multiplicity. Its two radicals can combine within the solvent cage to yield the combination product ethyl benzoate or the disproportionation products ethylene and benzoic acid (not shown). A certain fraction of the radicals will loose their singlet correlation because of dephasing, in which case they cannot combine or disproportionate but are more likely to attack the solvent and form the transfer product ethyl chloride by chlorine abstraction. Because of g (ethyl) < g (benzoyloxy) the g-value difference is negative, and because of a_H being negative for the CH_2 - protons in the ethyl radicals, the methylene protons in ethyl benzoate show emission, but in ethyl chloride they show enhanced absorption. The enhanced absorption found in the disproportionation product ethylene is due to the prevailing CH_3 - coupling in the ethyl radical, which has a positive a_H for the CH_3 - protons.

If shift reagents are added to the above system[9] in concentrations between 1 and 10% with respect to the peroxide concentration, the CIDNP of the ethyl benzoate is totally quenched, whereas the intensities of the other resonances are not significantly altered (figure 7.2). This means that the chelates, $Eu(fod)_3$ or $Pr(fod)_3$, for example[10] can be used to selectively remove resonances from CIDNP spectra. We explain this fact by assuming that the shift reagents complex specifically with products containing lone-pair electron atoms like O, N, or S, thereby shortening their spin-lattice relaxation times. As the spin-lattice relaxation times are the "memory" times of the CIDNP phenomenon which determine the intensities of the enhancements significantly, any reduction of T_1 results in quenching the CIDNP. However,

the fact that the intensities of the other products which
do not contain atoms with lone-pair electrons remain almost unchanged indicates that the presence of paramagnetic
shift reagents does not severely interfere with the nuclear
spin dependence of the intersystem-crossing step in free -
radical pairs.

Although adding shift reagents in such concentrations to
CIDNP - showing systems might be useful in simplifying the
spectra, information is lost when lines are removed totally. It is often more desirable to shift the lines just
enough apart to avoid overlapping. This can also be achieved by applying the shift reagents in lower concentrations.
Figure 8 shows the influence of $Pr(fod)_3$ (fig. 8.2) and
$Eu(fod)_3$ (fig. 8.3) on the spectra recorded during the decomposition of BPPO, if the molar ratio between the shift
reagent and the peroxide is kept as low as 7.5×10^{-3}. Now
the quartet of the CH_2 - protons is shifted upfield in the
presence of $Pr(fod)_3$ and downfield in the presence of Eu
$(fod)_3$, whereas the resonance lines of the other products
retain both their positions and their intensities.

The shifts achieved under such conditions are only of the
order of 0.1 ppm, but even these small changes can be very
important for special applications. Thus if different products give rise to coinciding resonances, their tendencies
to complex with a shift reagent can differ significantly.
This situation is demonstrated in figure 9, where we have
decomposed a mixture of BPPO and m-chlorobenzoyl propionyl
peroxide (CBPPO) simultanously in CCl_4 as the solvent.
Without shift reagent the quartets of the ethyl benzoate
and of the ethyl m-chlorobenzoate overlap and appear to be
one single product (fig. 9, lower trace). However, upon addition of $Pr(fod)_3$ to the system (fig. 9, upper trace) or
upon addition of $Eu(fod)_3$ (not shown), the ester resonance
separates into two emission quartets, which can easily be
analyzed separately. Apparently the chlorine-free ester
complexes better with the shift reagent than the chloro
isomer, hence its resonance lines are shifted more than
those of the ethyl m-chlorobenzoate.

In cases where emission and absorption lines overlap, the
application of shift reagents is even more rewarding. Thus
if ethyl benzoate is added to a solution of CBPPO in CCl_4

in such a quantity that during the reaction the intensities of the absorption quartet from the ethyl benzoate and of the emission quartet from the ethyl m-chlorobenzoate match, then no net emission or absorption remains, because of the exact overlap of the two quartets (fig. 10.1). If, however, $Eu(fod)_3$ is added to this system (fig. 10.2), the degeneracy of the emission and the absorption quartet is removed. Similarly, $Pr(fod)_3$ shifts the absorption quartet more upfield than the emission quartet (fig. 10.3).

Complications

There are certain problems encountered when applying shift reagents in systems which show the CIDNP phenomenon:
1.) Shift Reagent Poisons. Rather frequently in chemical reactions products are formed which react with the shift reagents. Often the shift reagents are destroyed, for example in the above system by the reaction product benzoic acid. If such poisons are formed in considerable quantities, a paramagnetic precipitate is formed, which reduces the CIDNP intensities of all products drastically, as has been shown for the BPO system in the presence of Fe^{3+} ions.
2.) Changing Concentration Ratios. During the reactions the products accumulate, which causes a change of the actual molar ratios of concentration between shift reagent and products. As a consequence of this time dependence, the line positions of the products which complex with the shift reagents move, and they tend to approach their chemical shift values without shift reagents. Unfortunately, the shifting lines restrict the uses of time-averaging methods. On the other hand the line positions reflect the conversion of the reactants during the chemical reaction. Thus, if the line positions are calibrated, they can be used to follow a reaction, and they allow one to determine the CIDNP - enhancement factors, which are difficult to obtain otherwise. The fate of the shift reagents can conveniently be studied by following the shifts of their own resonances during the reactions.
3.) Photochemical Reactions. The free-radical intermediates which cause the CIDNP phenomenon can also be generated photolytically. In this case the photochemical properties of the shift reagents have to be considered too. However, so far we have been unable to use shift reagents in systems where either free radicals or triplet states were generated photochemically. Thus we have irradiated various solutions

271

of aldehydes, ketones and peroxides inside the modified probe of a Varian HA 60 NMR spectrometer with the full arc of a 5 kW high pressure mercury-xenon lamp. Although we have obtained CIDNP spectra in all the above cases if shift reagents were omitted, no significant shifts of the lines could be achieved without quenching the intensities of all products so severely that they could not be observed any more. This failure appears to result from a variety of reasons. One problem is that the shift reagents themselves absorb light, particularly in the UV region. Furthermore, they themselves undergo photochemical changes, for example all chelates containing t-butyl group appear to split off t-butyl radicals, and give rise to CIDNP lines themselves. In addition the acetylacetonates of various heavy-metal ions are known to quench triplet states during photochemical reactions very efficiently. Accordingly we found a more drastic reduction of the CIDNP intensities in reactions which originated from excited triplet states than in those, where only excited singlet states were involved. Whatever the reasons for the failures are, it could well be that those shift reagents, which are not particularly useful in absorption NMR might well be superior to , for example,the Pr (fod)$_3$ or Eu(fod)$_3$ studied here. The evaluation of other shift reagents is still in progress, in particular as far as their behaviour in photochemically induced reactions is concerned.

In spite of some shortcomings, shift reagents can be used successfully in some CIDNP showing systems, and it can be expected that as more advanced shift reagents become available , their applicability in CIDNP will become more widespread.

References

1.) J. Bargon, Polymer Letters, 9, 681, (1971)
2.) J. Bargon, H. Fischer, and U. Johnsen, Z. Naturfschg. 22a, 1551, (1967)
3.) H.R. Ward and R. G. Lawler, J. Am. Chem. Soc. 89, 5518, (1967)
4.) J. Bargon and H. Fischer, Z. Naturfschg. 22a, 1556, (1967)
5.) G. L. Closs, J. Am. Chem. Soc. 91, 4552, (1969); G. L. Closs and A. D. Trifunac, ibid, 92, 2184, (1970); R. Kaptein and L. J. Oosterhoff, Chem. Phys. Lett. 4, 195, (1969)

6.) R. Kaptein, J. Chem. Soc. D , 732, (1971)
   G. L. Closs, Special Lectures presented at the XIIIrd
   IUPAC Congress, Volume 4, 19, (1971), Butterworth,
   (London)
7.) G.J. Hoytink, Acc. Chem. Res. 2, 114, (1969)
8.) R. A. Cooper, R. G. Lawler, and H. R. Ward, J. Am. Chem.
   Soc. 94,545, (1972)
9.) Some of the following results have appeared in preliminary form in: J. Bargon, J. Am. Chem. Soc. 95, 941,
   (1973).
10.) R. E. Rondeau, and R. E. Sievers, J. Am. Chem. Soc.
   93, 1522, (1971).

Figures 1 and 2 are adapted from reference 2; Figure 3
is adapted from reference 4. Figures 8, 9 and 10 are
taken from reference 9. The author thanks the copy-
right owners for the permission to reproduce the figures.

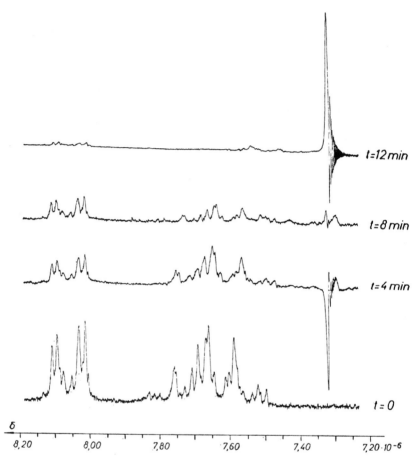

Figure 1: NMR spectra recorded during the thermal decomposition of a 0.05M solution of dibenzoyl peroxide in cyclohexanone at 110° C and a resonance frequency of 100 MHz.

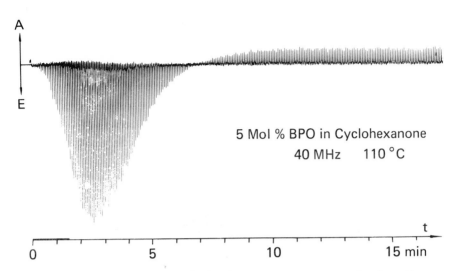

Figure 2: The changes of the benzene resonance during the thermal decomposition of dibenzoyl peroxide.

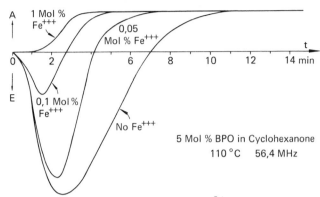

Figure 3: The influence of various Fe^{3+} concentrations on the emission intensity of the benzene resonance during the thermal decomposition of dibenzoyl peroxide.

Intersystem Crossing In Radical Pairs

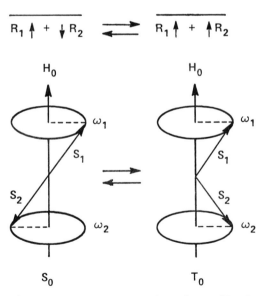

Figure 4: Intersystem crossing in radical pairs.

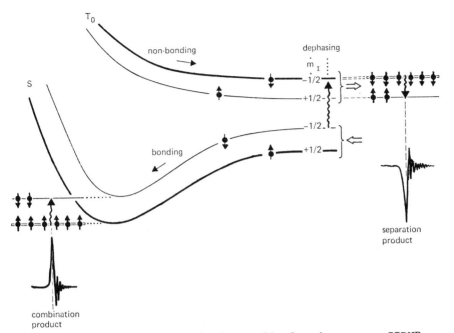

Figure 5: Dephasing in a singlet radical pair causes CIDNP
in the products. The schematic energy diagram
shows the consequences for nuclei in a radical
with the lower g-value and a positive hyperfine
coupling.

277

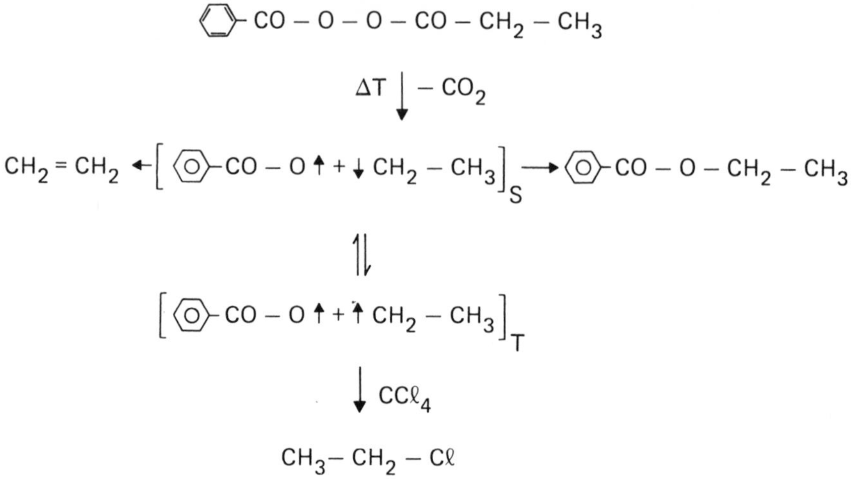

$$\langle\!\bigcirc\!\rangle\!-CO-O-O-CO-CH_2-CH_3$$

$$\Delta T \downarrow -CO_2$$

$$CH_2=CH_2 \leftarrow \left[\langle\!\bigcirc\!\rangle\!-CO-O\uparrow + \downarrow CH_2-CH_3 \right]_S \longrightarrow \langle\!\bigcirc\!\rangle\!-CO-O-CH_2-CH_3$$

$$\updownarrow$$

$$\left[\langle\!\bigcirc\!\rangle\!-CO-O\uparrow + \uparrow CH_2-CH_3 \right]_T$$

$$\downarrow CC\ell_4$$

$$CH_3-CH_2-C\ell$$

Figure 6: Decomposition mechanism of benzoyl propionyl peroxide in CCl₄.

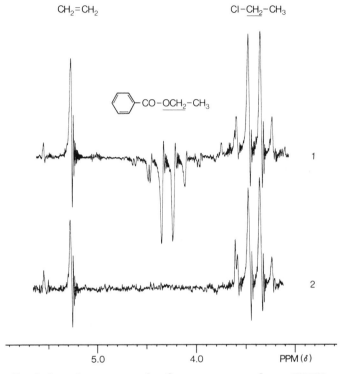

Figure 7: Selective removal of resonances from CIDNP spectra with shift reagents.

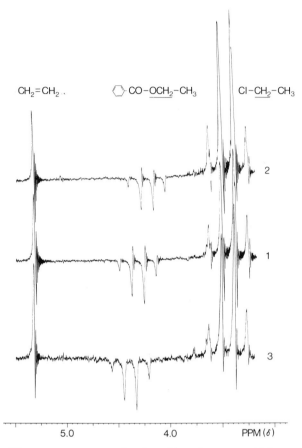

Figure 8: CH_2 -resonances during the decomposition of BPPO in CCl_4 at 60 MHz without shift reagent (1), with $Pr(fod)_3$ (2), and with $Eu(fod)_3$ (3).

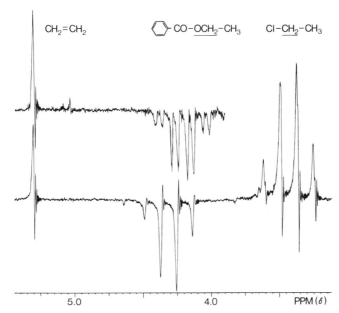

Figure 9: CH_2 - resonances during the decomposition of a mixture of BPPO and CBPPO in CCl_4 at 60 MHz with out shift reagents, and in presence of $Pr(fod)_3$ (upper trace).

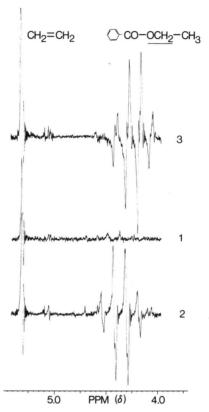

Figure 10: Emission quartet of ethyl benzoate and absorption quartet of ethyl m-chlorobenzoate without shift reagent (1), with Eu(fod)$_3$ (2), and with Pr(fod)$_3$ (3) at 60 MHz.

Ln(fod)$_3$ COMPLEXES AS NMR SHIFT REAGENTS:

States of Hydration; Self Association; Solution Adduct Formation and Changes of the NMR Time Scale.

C.S. Springer[*], A.H. Bruder, S. R. Tanny, M. Pickering and H.A. Rockefeller

Department of Chemistry, State University of New York at Stony Brook, Stony Brook, New York   11790

INTRODUCTION

The paramagnetic tris lanthanide (III) chelates of the anion of 6,6,7,7,8,8,8-heptafluoro-2,2-dimethyl-3,5-octanedione, H(fod), Ln(fod)$_3$, have been finding ever increasing usage as NMR shift reagents since the first report of Rondeau and Sievers.[1] The chemical property which permits this application is the Lewis acidity which the chelates possess as a consequence of their coordinative unsaturation. The neutral tris chelates dissolve in organic solvents and form labile adducts with a large variety of nucleophilic substrates. The paramagnetic lanthanide ions cause isotropic hyperfine shifts in the resonance frequencies of magnetic substrate nuclei.[2]

The Lewis acidity of these chelates also causes two side interactions which can interfere with their use as NMR shift reagents. These are: the formation of hydrates; and the formation of self-associated oligomers. We will discuss these problems in this chapter along with the nature of the adduct species present in solution.

Dynamic nuclear magnetic resonance (DNMR) spectroscopy has been extremely useful in the study of relatively rapid molecular motions and reactions.[3] A major problem which often plagues the application of this technique is the fixed nature of the "NMR time scale". The difference of the resonance frequencies of exchanging nuclei in the absence of exchange ($\Delta\nu_\infty$) is the main determinant of the magnitude of the "time scale". The size of $\Delta\nu_\infty$ is fixed by the magnetic properties of the environments between which the nuclei exchange. This limits the kinetic window,

i.e., the range of lifetimes (rate constants for first
order processes) which affect the spectral line shapes.
The problem usually arises because $\Delta\nu_\infty$ is (a) non-zero but
small or (b) zero because of accidental degeneracy. The
former situation means that the temperature or concentration conditions necessary to lengthen the lifetimes sufficiently are experimentally inaccessible or that only in-
accurate rate data can be obtained. The latter situation
means that no DNMR experiment is possible.

The selective nature of the lanthanide induced shift (LIS)
offers a method of overcoming the problem as manifest in
situations (a) and (b). In addition, the LIS provides for
a new DNMR technique. The conventional DNMR experiment
involves varying the rate(s) of the observed reaction(s)
(through changes in the temperature and/or concentrations
of reactants) while keeping the $\Delta\nu_\infty$ (or $\Delta\nu_\infty$'s) constant (by
operating at a constant H_O field strength). We will
demonstrate in this chapter how rate information can be
obtained by using the LIS to continuously vary $\Delta\nu_\infty$ through
the exchange sensitive region while keeping the rate of the
observed reaction constant (by operating at constant temperature and/or concentration).

STATES OF HYDRATION OF Ln(fod)$_3$ COMPLEXES

The characterization of the states of hydration of the
Ln(fod)$_3$ complexes is a difficult problem which still has
no entirely unequivocal solution.[4,5,6] At the time of the
initial synthesis of the chelates, we reported the existence of only the "monohydrated" state, Ln(fod)$_3 \cdot$OH$_2$, based
mainly upon our Karl Fischer titration and TGA evidence.[4,5]
Drying of the complexes by vacuum pumping at room temperature produced materials with no water titrable by the
Karl Fischer method and these were referred to as the
"anhydrous" species. The small amounts of water always detected by other techniques (IR and TGA) were attributed to
the hygroscopic nature of the "anhydrous" chelates.[4,5]
However, the structure of the praseodymium "monohydrate,
Pr(fod)$_3 \cdot$OH$_2$" determined by the x-ray crystallographic study
of de Villiers and Boeyens[7] has prompted a reinterpretation
of the previous results.[6]

De Villiers and Boeyens found that "Pr(fod)$_3 \cdot$OH$_2$" exists
in a dimeric form (shown schematically in I) in the solid.

Two β-diketonate anions   serve to bridge the metal ions in

$\underset{\sim}{\text{I}}$

a manner familiar for this type of ligand. A more significant aspect of this structure is that two kinds of "water"
oxygen atoms were found. One is located at appropriate
distances to serve as a third bridging atom between the
praseodymium ions. A second, which was much more difficult
to locate because of disorder and dehydration in the x-ray
beam, is in a position where the hydrogens could be hydrogen-bonded to the chelate oxygens and/or side chain
fluorines as shown in $\underset{\sim}{\text{I}}$. Further investigation has prompted
us to propose that only the hydrogen-bonded water is
titrable by the Karl Fischer technique.[6] Thus, the "mono-
hydrates" of the early lanthanide fod complexes are
probably more properly referred to as sesquihydrates,
$[(fod)_3Ln(OH_2)Ln(fod)_3] \cdot (OH_2)_2$. This is supported by a reinterpretation of the previously troublesome TGA data.[4,5,6]
The hydrogen-bonded water is only approximately stoichiometric. In a very damp atmosphere, even more water will be
taken up.[6]
   The TGA and IR evidence indicate that the bridging "water"
is not removed by vacuum pumping at room temperature; however, this "water" appears not titrable by the Karl Fischer
method.[6] (The bridging species may actually be a hydroxide
ion.) Thus, the early lanthanide complexes previously
referred to as "anhydrous" should probably be more properly
designated as hemihydrates, $[(fod)_3Ln(OH_2)Ln(fod)_3]$. The
truly anhydrous chelates of the early lanthanides seem to be
obtainable only by vacuum pumping at elevated temperature,[6]
or perhaps for prolonged periods at room temperature.

The above situation does not seem to apply to the later lanthanide complexes which show much reduced tendencies toward dimerization (<u>vide infra</u>). Thus, the monohydrated lutetium complex has a discrete monomeric seven-coordinate structure in the solid state[8] (II). The water molecule is

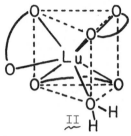

II

titrable by the Karl Fischer technique and can be removed by vacuum pumping at room temperature.

To summarize, the states of hydration of the $Ln(fod)_3$ chelates which seem to be attainable with some degree of reliability are the sesquihydrate (approximate) and the hemihydrate (approximate) for the early lanthanide ions; and the monohydrate for the later lanthanides. The Pr(III) and Eu(III) complexes, most used as NMR shift reagents, fall in the former category. The rigor necessary to ensure the use of truly anhydrous chelates for precise work cannot be overemphasized.

SOLUTION SELF ASSOCIATION AND ADDUCT FORMATION

Extensive studies of transition metal β-diketonate complexes have shown that whenever the coordination number of the metal can exceed twice its ionic charge, the neutral (inner) β-diketonate chelates can act as Lewis acids[9,10]. Such compounds are coordinatively unsaturated and attain higher coordination numbers by accepting Lewis bases as additional ligands. This acidity manifests itself in two general ways: self-association and adduct formation. These are exemplified in the chemistry of bis(acetylacetonate)nickel(II), $Ni(acac)_2$, some reactions of which are given in equations 1 and 2.[10] It can be seen that there is a strong drive toward six-coordination

four-coordinate   six-coordinate

$$3 \ Ni(acac)_2 \quad \xrightarrow{\text{cool}} \quad Ni_3(acac)_6 \qquad (1)$$

> 150°C, gas phase, or
non-coordinating
solvent

four-coordinate   five-coordinate   six-coordinate

$$Ni(acac)_2 \quad \xrightarrow{D} \quad Ni(acac)_2 D \quad \xrightarrow{D} \quad Ni(acac)_2 D_2 \qquad (2)$$

at the nickel ion. This is achieved by the formation of
new bonds to nucleophiles if possible or by sharing
chelate oxygens in a trimer if not. This type of behavior
is very common for the bis(β-diketonate) chelates of
divalent metal ions which can be five- or six-coordinate.[9,10]
   Since the ionic charge of the lanthanides in these
studies is three, a coordination number greater than six
is requisite for Lewis acidity. The abundance of known
eight-coordinate tetrakis (β-diketonate) lanthanide(III)
anions[2b] clearly demonstrates that the general phenomenon
of higher coordination numbers (i.e. > 6) for lanthanide
complexes is also important for the β-diketonates. Thus,
much of the chemistry of the simple tris fod complexes
can be understood in terms of their Lewis acidity.
   The self-association of Pr(fod)$_3$ and its hydrated forms
in CCl$_4$ is clearly revealed by the vapor pressure
osmometry data[6] shown in Figure 1. The average number of
monomers in a solute oligomeric molecule, $\bar{n}$, is plotted as
a function of the total concentration of monomeric units.
At a concentration of 0.01 M, all forms of the complex
have an $\bar{n} \geq 1.5$. It is clear, however, that decreasing
the state of hydration results in an increase in the extent of self-association. At 0.1 M, $\bar{n} \sim 2.0$ for Pr(fod)$_3$
while $\bar{n}$ is only ~ 1.7 for Pr(fod)$_3$(OH$_2$)$_{\sim 3/2}$. The solid
curves in Figure 1 represent the best least-squares fits
to the circled experimental data points found by a computer program which iterates the values of the equilibrium
quotients K$_2$ and K$_3$, defined in equations 3 and 4. The

$$2 \; \mathrm{Ln(fod)}_3 \mathrm{(OH_2)}_x \; \xrightleftharpoons{K_2} \; \left(\mathrm{Ln(fod)}_3 \mathrm{(OH_2)}_x \right)_2 \tag{3}$$

$$3 \; \mathrm{Ln(fod)}_3 \mathrm{(OH_2)}_x \; \xrightleftharpoons{K_3} \; \left(\mathrm{Ln(fod)}_3 \mathrm{(OH_2)}_x \right)_3 \tag{4}$$

values of K_2 and K_3 corresponding to the best theoretical curves are shown in the figure. Fits with $K_3 > 0$ for $\mathrm{Pr(fod)}_3 \mathrm{(OH_2)}_{\sim 3/2}$ were definitely poorer than that shown, as were fits with $K_3 = 0$ for $\mathrm{Pr(fod)}_3 \mathrm{(OH_2)}_{\sim 1/2}$ and $\mathrm{Pr(fod)}_3$.

These results lend themselves to an interesting interpretation in terms of a drive toward eight-coordination about the praseodymium ion. For any state of hydration equal to or above that of the hemihydrate, the extent of trimerization is small because the formation of a dimer with a bridging "water" molecule, in which both metal ions become eight-coordinate, is possible. This is shown in equation 5 for $\mathrm{Pr(fod)}_3 \mathrm{(OH_2)}_{\sim 1/2}$.

six-coordinate   seven-coordinate

$$\mathrm{Pr(fod)}_3 \quad + \quad \mathrm{Pr(fod)}_3 \mathrm{(OH_2)} \; \rightleftharpoons \tag{5}$$

eight-coordinate

$$\left[\mathrm{(fod)}_2 \mathrm{Pr(fod)}_2 \mathrm{(H_2O)Pr(fod)}_2 \right]$$

structure similar to that in I

However, if the chelate is completely anhydrous, a reasonable analysis indicates that the metal ions cannot both attain eight-coordination by simple dimerization. The lowest oligomer in which all metal ions can become eight-coordinate is the trimer.[6] This is shown in equation 6.

six-coordinate   seven-coordinate

$$2 \; \mathrm{Pr(fod)}_3 \; \rightleftharpoons \; \mathrm{Pr_2(fod)}_6 \; \xrightarrow{\mathrm{Pr(fod)}_3} \tag{6}$$

structure similar
to III

eight-coordinate
$\mathrm{Pr_3(fod)}_9$
structure similar to IV

The structure of the intermediate dimer is probably similar
to that found for $Pr_2(thd)_6$[11] (thd = 2,2,6,6-tetramethyl-
3,5-heptanedionate, sometimes referred to as dpm) and
shown schematically in III while the structure of the

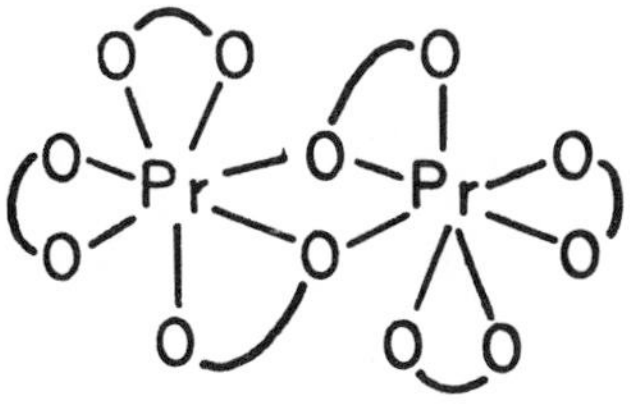

III

trimer is predicted to be similar to that shown schematically in IV.

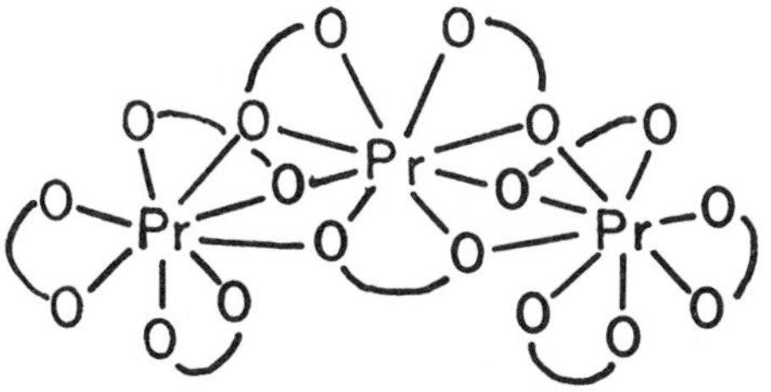

IV

   Similar results were obtained for $Eu(fod)_3$ and its hydrated forms and the best fit association quotients along
with those of the $Pr(fod)_3$ complexes are set out in
Table I. In the table, K_3 is replaced by K_3', defined in
equation 7, and calculated from the relationship

$$(Ln(fod)_3(OH_2)_x)_2 + Ln(fod)_3(OH_2)_x \xrightleftharpoons{K_3'}$$

$$(Ln(fod)_3(OH_2)_x)_3$$

(7)

289

$K_3' = K_3/K_2$. The results obtained, in similar studies of "Pr(fod)$_3$" and "Eu(fod)$_3$" in several solvents, by Desreux, Fox and Reilley[12] are also included in Table I. The chelate formulae are placed in quotation marks because it is not clear exactly how the complexes were dried immediately prior to the osmometry studies. Inspection of the data presented in Table I allows us to enumerate three factors which affect the extent of self-association of Ln(β-diketonate)$_3$ chelates. One is the state of hydration as has been discussed above. Another is the polarity of the solvent. The data of Desreux, Fox and Reilley indicate that the extent of self-association has the following solvent dependence: n-C$_6$H$_{14}$ > CCℓ_4 > CHCℓ_3. (See reference 13 for recent results in benzene.) That is, self-association is reduced in the more polar solvents. This is an important point because the most popular solvents for shift reagent studies have been CHCℓ_3, where self-association is very small, and CCℓ_4, where it is quite significant.

The third factor apparent from Table I is the effect of the size/acidity of the metal ion. The chelates of the larger praseodymium ion are more highly associated than those of the europium ion under the same conditions. This is in agreement with the general trend across the entire lanthanide series as exemplified by the (+)-3-trifluoro-acetylcamphorates[14] and the tropolonates[15] and is also substantiated by more recent self-association studies reported in this book.[13,16] Although the metal ions become more acidic as they get smaller (and therefore more "highly" coordinatively unsaturated), the steric requirements for the hindered process of self-association become more detrimental and apparently override the increased acidity. For the later lanthanide β-diketonates, coordination numbers above seven are rare.[2b]

A fourth factor is apparent upon consideration of recent osmometric studies of the Ln(thd)$_3$ complexes.[12,17,18] All of these studies indicate the monomeric nature of the chelates in all solvents and at all concentrations studied. These results are consistent with the observed monomeric structures of the bis thd complexes of the divalent first row transition metal ions.[10] (In fact, the only instance of known self-association of a thd complex is the solid state Pr$_2$(thd)$_6$ dimer (III).) Graddon has pointed out that it is not clear whether the monomeric character of thd

chelates is derived from the bulky nature of the t Bu substituents, as has often been supposed, or the great
basicity of the thd anion[10] (the pK_a of H(thd) is 15.9).
The latter factor would serve to lessen the acidic character of the metal ion and thus reduce its tendency toward
coordinative expansion. This seems to be an important
factor in Ni(II) β-diketonate chemistry[10] and most likely
also for the Ln(III) complexes. The H(fod) ligand (pK_a =
8.7) is, after all, not a great deal less sterically
hindered than the H(thd) ligand. Recent proposals in the
shift reagent literature, based on NMR data, of the dimeric
nature of Ln(thd)$_3$ complexes[12] in solution are almost certainly incorrect.

The acidity of the lanthanide β-diketonates is, of course,
also exhibited by the formation of labile solution adducts
(the active shift reagent species) with nucleophilic
substrates. We have investigated the interaction of tri-
methyl carbamate (TMC), $\underset{\sim}{V}$ with Eu(fod)$_3$(OH$_2$)$_{\sim 1/2}$ and

$$H_3CO \quad C \quad O$$
$$H_3C - N - CH_3$$

trans   cis

$\underset{\sim}{V}$

Eu(fod)$_3$ in solution.[6] TMC is an ideal substrate for solution adduct formation studies and for the DNMR studies to
be described below.

The plots of the isotropic hyperfine shifts[2], Δ, of the
various substrate proton magnetic resonance frequencies
as a function of the mole ratio, ρ, (chelate/substrate) are
shown in Figure 2. The shifts of all resonances are to
lower fields, as is usually true with europium (III)[2], and
as has been previously reported for carbamates.[19,20] The
OCH$_3$ group resonance undergoes the largest downfield shift
in the presence of the paramagnetic Eu(III) complex. This
is consistent with coordination through either the carbonyl
or the ester oxygen atoms. Carbonyl groups usually show a
greater interaction with Ln(β-diketonate)$_3$ chelates than

sterically crowded ethers.[2b] Coordination through the carbonyl oxygen is confirmed by the fact that, while TMC interacts with $Eu(thd)_3$, trimethyl thiocarbamate $((H_3C)_2 NC(S)OCH_3)$ does not.[19] The NCH_3 resonance shifted farthest downfield is assigned to the <u>cis</u> NCH_3 group because of its closer proximity to the site of complexation.[20]

The solid curves through the $Eu(fod)_3$ points (circles) in Figure 2 are the best least-squares fits obtained by assuming the four competing equilibria shown in Figure 2 to be the important ones.* The values of K_2 and K_3' for the dimerization and trimerization, respectively, of $Eu(fod)_3$ are those obtained from the osmometry studies described above. These parameters are simply constants which regulate the amount of monomeric $Eu(fod)_3$ present. Thus, the fitting problem reduces to one of four variable parameters: K^{AD} and K^{AD2}, which are defined in equations 8 and 9; and Δ_1^O and Δ_2^O, the shifts of the substrate

$$A + D \underset{\xrightarrow{\hspace{1cm}}}{\overset{K^{AD}}{\rightleftharpoons}} AD \qquad (8)$$

$$AD + D \underset{\xrightarrow{\hspace{1cm}}}{\overset{K^{AD2}}{\rightleftharpoons}} AD_2 \qquad (9)$$

resonances in the 1:1 and 1:2 complexes, respectively. The best fit values of K^{AD} and K^{AD2} are; $16 \pm 6 \times 10^2$ (M^{-1}) and 107 ± 10 M^{-1}, respectively and are shown in Figure 2. The best fit values of Δ_1^O and Δ_2^O for each resonance are given in Table II.

The importance of attaining eight-coordination about the metal ion in the early lanthanide β-diketonates, evidenced

---

*Other possible reactions, $A_2 + D \rightleftharpoons A_2D$ and $A_2D + D \rightleftharpoons A_2D_2$, were ignored because they should be less important (especially at low values of ρ) and their inclusion would introduce two new parameters each. An eight parameter fit to this data would not be of significance proportional to its difficulty. These reactions almost certainly become important at larger values of ρ (> 1.5) and must be taken into account in that region.

---

in the self-association results described above, is also manifest in the formation of 1:2 adducts (AD_2) with mono-dentate substrates in solution. Although much of the early shift reagent work with the $Ln(fod)_3$ chelates assumed the formation of only 1:1 adducts,[2] the appearance of the mole ratio plots for the interaction of the small nucleophilic substrate TMC with anhydrous $Eu(fod)_3$ leaves no doubt as to the formation of 1:2 adducts. The curves begin to level off at values of $\rho \cong 0.5$. This indicates appreciable amounts of 1:2 adduct formation at lower values of ρ.[2b]

Recently published mole ratio plots also indicate the existence of 1:2 adducts of $Eu(fod)_3$[21] and $Pr(fod)_3$[22] in solution as do others presented in this book.[13,23] Although these curves do not begin to level off until $\rho \cong 1.0$, careful analysis reveals the presence of signifi-cant amounts of the 1:2 adducts. The delayed leveling off is due to the fact that Δ_1^o is greater than Δ_2^o; a result which seems to be the most common. For the carbamate studies presented here, however, Δ_2^o is greater than Δ_1^o (Table II). In the mole ratio plots of systems containing 1:2 adducts, the initial slopes are equal to $2\Delta_2^o$.[2b,21,24] That is, in the low ρ region where AD_2 adducts predominate, the observed shift is tending toward Δ_2^o, where it would level off at $\rho = 0.5$ if 1:2 complexes were exclusively formed. However, as ρ increases, the AD species become more important and, when the Δ_1^o values are smaller than the Δ_2^o values, the curves begin to level off at values of Δ intermediate between Δ_2^o and Δ_1^o, and therefore well below Δ_2^o.*

Further evidence for the formation of 1:2 adducts is found from linear Scott-type NMR plots of substrate binding to $Eu(fod)_3$,[25] the low temperature NMR spectrum of a 1:2 adduct of $Eu(fod)_3$ with DMSO,[26] and a circular dichroism study of $Eu(fod)_3$ adducts in solution.[27] There is also evidence from mole ratio plots that Eu(III) chelates of other fluorinated β-diketonates; fhd(1,1,1,2,2,6,6,7,7,7-decafluoro-3,5-heptanedionate),[28] and pta(1,1,1-trifluoro-

---

*This is a significant observation because the values of Δ_1^o have often been estimated from the initial slopes of mole ratio plots in shift reagent studies. Also, estimates of Δ_1^o values have recently been made from the values of Δ on the level portion of the mole ratio plot.[22] The mole ratio plot is by far the most common type used in shift reagent studies.

5,5-dimethyl-2,4-hexanedionate)[24] form 1:2 adducts in
solution. The latter curves begin to level off or go
through maxima at $\rho \cong 0.5$, presumably for the same reason
as those of the carbamate. Reuben has presented other
examples in this book where Δ_2^o is greater than Δ_1^o for
Eu(fod)$_3$ adducts.[16] Marks, et al, have presented data
indicating that the tris fod chelate of the larger Pr(III)
ion can form 1:3 adducts.[13]

Although a number of crystal structures have been determined which have eight-coordinate 1:2 adducts of Ln(thd)$_3$
chelates with monodentate substrates,[2b] no direct evidence
has been noted of solution 1:2 adduct formation of these
chelates of non-fluorinated ligands. This might be
expected to be less important if the high basicity of the
thd anion is a dominant factor as discussed above for
self-association.

The value of K^{AD} = 1600 M^{-1} obtained for the TMC adduct
is considerably larger than the few reliable numbers found
in the literature for 1:1 adducts of Eu(fod)$_3$. These are
all reported to be between 222 and 280 M^{-1}[6] (see, however,
refs. 16 and 23). The larger number is due to the fact
that we have taken into account the competition of self-
association. A value of K^{AD} = 375 M^{-1} was obtained by
fitting the cis NCH$_3$ curve of the interaction of anhydrous
Eu(fod)$_3$ with TMC with the assumption of $K_2 = K_3 = 0$.[6]
Thus, the tris fod chelates are quite respectable Lewis
acids. The non-fluorinated thd chelates are less acidic
as would be expected.[2b] A considerable solvent effect on
the values of the association quotients has been noted.[29]

The effects of water, in the form of the hemihydrated
Eu(fod)$_3$, are also evident in Figure 2. The dashed curves
through the Eu(fod)$_3$(OH$_2$)$_{\sim 1/2}$ data (triangles) are not
computer fitted. They merely emphasize the effect of water
on the shift reagent plots. A pronounced non-monotonicity
is clear in the curves of the OCH$_3$ and cis NCH$_3$ resonances
which go through maxima at $\rho \cong 0.67$. Maxima in mole ratio
plots which bear some resemblance to these have been reported in the literature.[21,24,30] These are all for the
interaction of substrates with chelates of fluorinated
ligands.

For the data presented here, the most likely source of
the nonmonotonicities is the presence in solution of the
adduct Eu(fod)$_3$(OH$_2$)(TMC). When Eu(fod)$_3$(OH$_2$)$_{\sim 1/2}$ is used
as the acid, the values of Δ continue to rise past the

point where they leveled off for the anhydrous chelate. They go through a maximum at $\rho \cong 0.67$ before declining back towards the Δ_1^o values. This would be consistent with a value of Δ^o for the $Eu(fod)_3(OH_2)(TMC)$ complex larger than Δ_2^o. The fact that the curves go through a maximum at $\rho \cong 0.67$ may indicate the importance of the reaction given in equation 10 in which all europium ions become eight-

$$Eu(fod)_3 + Eu(fod)_3(OH_2) + 3TMC \quad \rightleftharpoons$$

(10)

$$Eu(fod)_3(TMC)_2 + Eu(fod)_3(OH_2)(TMC)$$

coordinate. At values of ρ greater than 0.67, some seven-coordinate $Eu(fod)_3(TMC)$ must be formed.

Use of $Eu(fod)_3(OH_2)_x$, where $x > 3/2$, as a shift reagent results in mole ratio curves which are generally displaced to lower values of Δ, for a given ρ, than those shown in Figure 2.[6] Similar effects have been reported by Marks, et al,[13] and Sievers and coworkers[31] in this book.

CHANGING THE NMR TIME SCALE

Trimethylcarbamate ($\underset{\sim}{V}$) is a substrate particularly well suited to illustrate and test the new DNMR technique described in the INTRODUCTION. It is known that there is hindered rotation about the carbonyl-nitrogen bond; the non-equivalence of the <u>cis</u> and <u>trans</u> NCH_3 resonances has been observed in 10 mole % solution in $CDCl_3$ below $-23°C$.[32,33] However, the $\Delta\nu_\infty$ observed at these temperatures is quite small (0.030 ppm,[32] 0.032 ppm[33]) thereby reducing the precision of the activation parameters, derived from a traditional DNMR study.[32] In 25% solutions in CH_2Cl_2 or toluene, no splitting of the NCH_3 resonance is seen, even at $-46°C$ (60 MHz).[34] The NCH_3 resonances are also isochronous at ambient probe temperature in CCl_4 solution at 60 MHz.[34,35] Situation (b), described in the INTRODUCTION, obtains in this case. That is, $\Delta\nu_\infty$ is zero because of accidental degeneracy.[36]

Figure 3 depicts examples of the proton NMR spectrum of TMC in CCl_4 for different values of the mole ratio (ρ) of $Eu(fod)_3(OH_2)_x$ (where $x < 1/2$). The NCH_3 resonance, which

is a single sharp peak with twice the area of the OCH_3
resonance in the diamagnetic CCl_4 solution[34,35] ($\rho = 0$), is
broadened and then split into two peaks as more chelate is
added. The most important observation is that the NCH_3
resonances resharpen after their splitting, and after ρ
reaches approximately 0.5, three equally sharp resonances
are observed. This is a clear indication that chemical
exchange is affecting the shape of the resonance lines.
A smaller, steady increase in linewidth of all peaks with
ρ is caused by the paramagnetic ion. This is more clearly
seen in Figure 4 where the linewidths (full width at half-
height) of all the resonances are plotted as a function of
ρ. The OCH_3 group, which undergoes no chemical exchange,
shows only the broadening effect of the unpaired metal 4f
electrons.* The NCH_3 resonances, on the other hand, show
a tremendous increase in broadening due to the intramolecular exchange of the $N-CH_3$ groups between the cis and trans
environments. As $\Delta \nu$ increases toward its value in the
$Eu(fod)_3(TMC)_2$ adduct ($\Delta\Delta_2^o = 3.01$ ppm, Table II), however,
the exchange has less of an effect on the linewidth which
therefore decreases back to that of the OCH_3 resonance.
This behavior implies that kinetic information on the
process causing the broadening can be obtained from the
data gathered at constant temperature and/or concentration
of substrate.

We have used a computer program, based on the classical
modified Bloch equations for uncoupled two-site exchange,[3b]
to obtain observed rate constants from the total lineshape
analysis of the $N-CH_3$ resonances as a function of ρ.[36] This
program requires, as input parameters, the linewidths
(T_2 values) and the separation ($\Delta \nu_\infty$) of the exchanging
resonances in the absence of exchange. The linewidth of
the non-exchanging OCH_3 resonance at each value of ρ was
used to calculate the value of T_2 for the NCH_3 resonances
in the absence of exchange at that value of ρ. For almost
all of the range of ρ values studied, the exchanging peaks
were far enough apart that the observed separation $\Delta \nu$ was
equal to $\Delta \nu_\infty$ (a distinct advantage of this method). Only
for the very smallest mole ratios just before and, of

---

*The slight maximum in the OCH_3 curve at $\rho \cong 0.25$ may be a
 manifestation of the intermolecular exchange (of free and
 complexed TMC) term in the expression for the transverse
 relaxation time.[37]

course, just after coalescence ($0.026 < \rho_c < 0.051$, 27°C)
were estimates of $\Delta\nu_\infty$ necessary. These were made by extra-
polation of a $\Delta\nu$ versus ρ plot.

Some of the computer-fitted spectra are shown in
Figure 5. As can be seen, the fitting is quite good. The
pre-exchange lifetimes, τ_{obs}, obtained from these fittings
are shown in the figure. The first-order rate constants
are related to the τ_{obs} by the equation: $k_{obs} = 1/2\ \tau_{obs}$.

Figure 6 shows a plot of k_{obs} versus σ, the saturation
fraction. The saturation fraction is the fraction of TMC
substrate bound to Eu(fod)$_3$ and is given by equation 11.

$$\sigma = \frac{[AD]_{eq} + 2[AD_2]_{eq}}{[D]_{eq} + [AD]_{eq} + 2[AD_2]_{eq}} \tag{11}$$

It was calculated for each value of ρ by using the association quotients given in Figure 2. The plot is linear over
90% of the saturation fraction, from 0.1 to 1.0, with a
small but non-zero slope. The k_{obs} values obtained for
$\sigma < 0.1$ are not accurate because the peaks had coalesced
and a good value of $\Delta\nu_\infty$ could not be reliably estimated.
Thus, it is possible to obtain kinetic information on the
constant rate process by <u>varying the NMR time scale through
the exchange-sensitive region</u>. This is done by changing
$\Delta\nu_\infty$ from zero (obtained by extrapolating $\Delta\nu$ to ρ = 0)[20] to
3.01 ppm through the operation of the LIS. Since the lin-
ear portion of the curve in Figure 6 has a non-zero slope,
one must ask whether or not the rate of the process
observed is truly constant. The answer is yes, with the
qualification that the observed rate is the average rate
of <u>two processes</u> whose relative importances are changing
with σ.

The weak TMC: Eu(fod)$_3$ adduct is extremely labile.[2b] The
lifetime of the quinoline: Eu(thd)$_3$ adduct under similar
conditions has been estimated to be 210 nsec.[37] Although
the lifetime of the stronger Eu(fod)$_3$ adduct is expected
to be longer, it is still much shorter than the isomerization lifetime which is on the order of 40 msec. The intermolecular exchange is certainly fast on either of the two
NMR time scales. Thus, it seems clear that the observed
rate constant is the weighted average of the rate constant

for isomerization of free TMC, k_{free}, and that for complexed TMC, k_{adduct} (equation 12).

$$k_{obs} = (1-\sigma)\, k_{free} + \sigma k_{adduct} \qquad\qquad (12)$$

The value of k_{free} can be found by extrapolating the linear portion of the curve to $\sigma = 0$, to be 33.0 ± 3.7 sec^{-1} at 27°C in CCℓ_4. Employing the standard Eyring equation (with the usual assumption of the transmission coefficient being unity), the free energy of activation for isomerization of uncomplexed TMC, $\Delta G^{\neq}_{300}$, is calculated to be 15.5 ± 0.1 kcal/mole. This is in excellent agreement with the value of 15.2 kcal/mole obtained by Inglefield and Kaplan from their total lineshape analysis of the DNMR spectrum of TMC in CDCℓ_3 solution at temperatures near the coalescence (-13°C).[32] The values of $\Delta G^{\neq}$ for this type of isomerization have been found to be almost temperature and, surprisingly, solvent independent.[3] One might have expected the CCℓ_4 value to be somewhat smaller than the CDCℓ_3 value, but it seems that the main DNMR problem in CCℓ_4 is caused by the reduction of $\Delta\nu_\infty$. The only other values of $\Delta G^{\neq}$ for isomerization of N,N-dimethylcarbamates, have been obtained by DNMR methods more approximate than that of Inglefield and Kaplan. Values of 14.4 and 14.5 have been obtained for the O-phenyl and O-naphthyl derivatives respectively.[38] A value of 15.9 has been obtained for the O-benzyl derivative in both CDCℓ_3 and pyridine.[39]

Thus, we have demonstrated and successfully tested a method of determining rate constants and activation parameters for processes which cannot be studied under normal, diamagnetic, NMR conditions through the use of the paramagnetic LSR. In this case, in the diamagnetic CCℓ_4 solution, $\Delta\nu_\infty$ is zero at 60 MHz: isomerization cannot be studied at any temperature. The remaining activation parameters could be determined by repeating the studies reported above for several different temperatures.

Extrapolating the k_{obs} vs. σ curve to $\sigma = 1$, gives a rate constant for isomerization of complexed TMC, k_{adduct}, of 0.5 ± 3.7 sec^{-1} at 27°C (see Figure 6). A rate constant of 0.5 sec^{-1} corresponds to a $\Delta G^{\neq}$ value of 18.0 kcal/mole at 27°C. However, k_{adduct} is zero within experimental error. The only definite statement which can be made is that complexation of TMC to $Eu(fod)_3(OH_2)_x$ retards the process of isomerization. This is to be expected for both electronic and steric reasons. Coordination through the carbonyl oxygen would lead to increased double bond character in the carbonyl carbon-nitrogen bond due to relative stabilization of resonance form VII. Protonation of amides at the carbonyl oxygen has been found to increase the

$$H_3CO \quad O \qquad \longleftrightarrow \qquad H_3CO \quad O^- $$

VI   VII

barrier to isomerization by 0.4 kcal/mole.[40] Substrate molecules have been found to be quite crowded in the LSR adducts which have been studied by x-ray crystallography.[2b] This would almost certainly lead to some steric hindrance to substrate isomerization in the adduct complex. Therefore, an increase in the activation free energy seems quite plausible.

Figure 7 shows the spectrum of an azepine carbamate as a function of the mole ratio of $Eu(fod)_3$. A splitting and kinetic broadening of the methylene resonances can be clearly observed. This system has not yet been treated quantitatively.

This technique should prove quite useful for studying the kinetics of previously unobservable processes because the LSR have been found to interact with a wide range of Lewis basic substrates.[2] A particularly interesting area of application may be the study of fluxional organometallic molecules. Marks and coworkers[13] and Foreman and Leppard[41] have shown that the LSR interact with a large variety of coordinated ligands. Kinetic studies of rapid intermolecular reactions should also be amenable to this method.

It is interesting to speculate on the limits of the magnitudes of rate constants which can be studied by this technique. Initial slopes of mole ratio plots corresponding to eventual values of $\Delta\nu = 5 \times 10^4$ Hz, have been reported.[24,42]

Use of equation 13,[3b] gives a value of the rate constant at

$$k_{coal.} = \pi\Delta\nu/\sqrt{2} \tag{13}$$

coalescence, $k_{coal.}$, of $\sim 10^5$ sec^{-1}. This corresponds to a free energy of activation, $\Delta G^{\ddagger}_{300}$, of 10.7 kcal/mole at 27°C. However, the <u>largest</u> values of $\Delta\nu$ to be expected in practical cases are probably closer to 5×10^3 Hz. This would correspond to a $k_{coal.}$ of $\sim 10^4$ sec^{-1} and a free energy of activation $\Delta G^{\ddagger}_{300}$ of ~ 12 kcal/mole.

ACKNOWLEDGEMENTS

The authors wish to thank Professor F. W. Fowler of Stony Brook for many stimulating and helpful discussions concerning the DNMR work. MP(PRF Postdoctoral Fellow) and CSS wish to thank the donors of the Petroleum Research Fund, administered by the American Chemical Society, for support of this work (Grant #5672 AC14). SRT wishes to thank the National Science Foundation (Grant #GP20099) for financial support during the period of this research.

References

1. R. E. Rondeau and R. E. Sievers, Jour. Amer. Chem. Soc.
    93, 1522 (1971).
2. (a) R. von Ammon and R. D. Fischer, Angew. Chem. Inter.
    Ed. 11, 675 (1972); (b) R. E. Sievers, M. F. Richardson
    and C. S. Springer, review of lanthanide β-diketonates,
    to be published.
3. (a) H. Kessler, Angew. Chem. Inter. Ed. 9, 219 (1970);
    (b) G. Binsch, Topics in Stereochem. 3, 97 (1968).
4. C. S. Springer, D. W. Meek and R. E. Sievers, Inorg.
    Chem. 6, 1105 (1967).
5. R. E. Sievers, K. J. Eisentraut, C. S. Springer and
    D. W. Meek, Adv. Chem. 71, 141 (1968).
6. A. H. Bruder, S. R. Tanny, H. A. Rockefeller and
    C. S. Springer, submitted for publication.
7. J. P. R. de Villiers and J. C. A. Boeyens, Acta Cryst.
    B27, 692 (1971).
8. J. C. A. Boeyens and J. P. R. de Villiers, Jour. Crys.
    Mol. Struct. 1, 297 (1971).
9. J. P. Fackler, Prog. Inorg. Chem., 7, 361 (1966).
10. D. P. Graddon, Coord. Chem. Rev. 4, 1 (1969).
11. C. S. Erasmus and J. C. A. Boeyens, Acta Cryst. B26,
    1843 (1970).
12. J. F. Desreux, L. E. Fox and C. N. Reilley, Anal. Chem.
    44, 2217 (1972).
13. T. J. Marks, R. Porter, J. S. Kristoff and D. F. Shriver
    Chapter 12.
14. B. Feibush, M. F. Richardson, R. E. Sievers, and
    C. S. Springer, Jour Amer. Chem. Soc., 94, 6717 (1972).
15. E. L. Muetterties and C. M. Wright, ibid., 88, 4856
    (1966).
16. J. Reuben, Chapter 16.
17. V. A. Mode, and G. S. Smith, Jour. Inorg. Nucl. Chem.
    31, 1857 (1969).
18. J. S. Ghotra, F. A. Hart, G. P. Moss and M. L. Staniforth
    J. C. S. Chem. Comm. 113 (1973).
19. R. A. Bauman, Tet. Lett. 419 (1971).
20. L. R. Isbrandt and M. T. Rogers, Chem. Commun. 1378
    (1971).
21. B. L. Shapiro and M. D. Johnston, Jour. Am. Chem. Soc.
    94, 8185 (1972).
22. J. W. ApSimon, H. Beierbeck, A. Fruchier, ibid. 95,

939 (1973).

23. B. L. Shapiro, M. D. Johnston, A. D. Godwin, H. L. Pearce, T. W. Proulx, M. J. Shapiro and F. A. Reilly, Chapter 11.

24. J. K. M. Sanders, S. W. Hanson and D. H. Williams, Jour. Am. Chem. Soc., 94, 5325 (1972).

25. V. G. Gibb, I. M. Armitage, L. D. Hall and A. G. Marshall, ibid. 94, 8919 (1972).

26. D. F. Evans and M. Wyatt, J. C. S. Chem. Commun. 312 (1972).

27. N. H. Andersen, B. J. Bottino and S. E. Smith, ibid. 1193 (1972).

28. C. A. Burgett and P. Warner, J. Mag. Res., 8, 87 (1972).

29. J. Bouquant and J. Chuche, Tet. Lett. 493 (1973).

30. K. Roth, M. Grosse and D. Rewicki, ibid. 435 (1972).

31. D. S. Dyer, J. A. Cunningham, J. J. Brooks, R. E. Sievers and R. E. Rondeau, Chapter 2.

32. P. T. Ingelfield and S. Kaplan, Can. Jour. Chem. 50, 1594 (1972).

33. E. Lustig, W. R. Benson and N. Duy, Jour. Org. Chem. 32, 851 (1967).

34. R. L. Middaugh, R. S. Drago and R. J. Niedzielski, Jour. Am. Chem. Soc. 86, 388 (1964).

35. A. E. Lemire and J. C. Thompson, Can. Jour. Chem. 48 824 (1970).

36. S. R. Tanny, M. Pickering and C. S. Springer, submitted for publication.

37. J. Reuben and J. S. Leigh, Jour. Amer. Chem. Soc. 94, 2789 (1972).

38. V. Machacek and M. Vecera, Coll. Czech. Chem. Comm. 37, 2928 (1972).

39. B. J. Price, R. V. Smallman and I. O. Sutherland, Chem. Comm., 319 (1966).

40. G. Fraenkel and C. Franconi, Jour. Amer. Chem. Soc. 82, 4478 (1960).

41. M. I. Foreman and D. G. Leppard, Jour. Organomet. Chem. 31, C31 (1971).

42. C. Beaute, Z. W. Wolkowski and N. Thoai, Tet. Lett. 817, 821 (1971).

Table I.   VAPOR PRESSURE OSMOMETRY STUDIES

Compound	Solvent						Reference
	$CHCl_3$ (37°C)		CCl_4 (37°C)		$n\text{-}C_6H_{14}$		
	$K_2(M^{-1})$	$K_3'(M^{-1})$	$K_2(M^{-1})$	$K_3'(M^{-1})$	$K_2(M^{-1})$	$K_3'(M^{-1})$	
$Pr(fod)_3(OH_2)_{\sim 3/2}$	–	–	286±20	0	–	–	this work
$Pr(fod)_3(OH_2)_{\sim 1/2}$	–	–	314±20	15±2	–	–	"
$Pr(fod)_3$	–	–	140±8	45±5	–	–	"
"$Pr(fod)_3$"	1.6	0	339	0	5130	224	12
$Eu(fod)_3(OH_2)_{\sim 3/2}$	–	–	75±5	0	–	–	this work
$Eu(fod)_3(OH_2)_{\sim 1/2}$	–	–	68±5	6.9±1	–	–	"
$Eu(fod)_3$	–	–	367±22	12±2	–	–	"
"$Eu(fod)_3$"	1.6	0	100	0	5010	12	12

Table II.   NMR SHIFT PARAMETERS FOR THE INTERACTION
OF TMC WITH $Eu(fod)_3$

Resonance	Δ_1^o (ppm)	Δ_2^o (ppm)
OCH_3	$6.94 \pm .03$	$9.65 \pm .01$
cis-NCH_3	$6.32 \pm .02$	$7.49 \pm .01$
trans-NCH_3	$4.14 \pm .03$	$4.48 \pm .01$

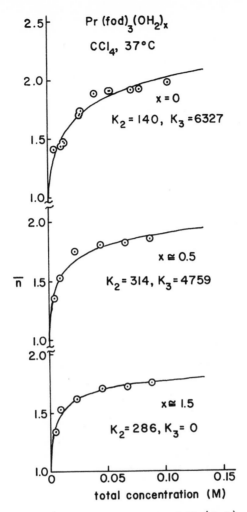

Figure 1. Self-association curves of $Pr(fod)_3$, $Pr(fod)_3-$ $(OH_2)_{\sim 1/2}$, and $Pr(fod)_3(OH_2)_{\sim 3/2}$ as a function of concentration in CCl_4 at $37^\circ C$. $\bar{n}$ is the average number of monomers per solute molecule (determined from vapor pressure osmometry). The total concentration is calculated assuming that all solute molecules are monomers. The circles represent experimental data points. The lines are theoretical curves obtained with the values of the association quotients for dimerization and trimerization (K_2 and K_3) shown.

305

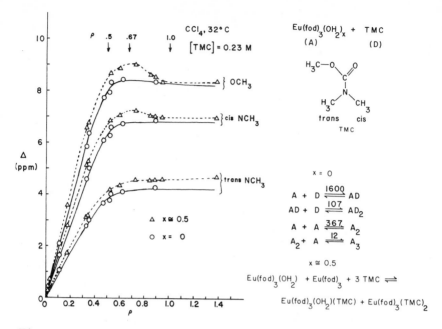

Figure 2. Mole ratio plots of the isotropic hyperfine shifts of the resonances of trimethyl carbamate induced by $Eu(fod)_3$ and $Eu(fod)_3(OH_2)_{\sim 1/2}$ in CCl_4 at 32°C. The circles and triangles represent experimental data points. The solid lines represent the theoretical curves obtained with the association quotients shown for the $Eu(fod)_3$ equilibria.

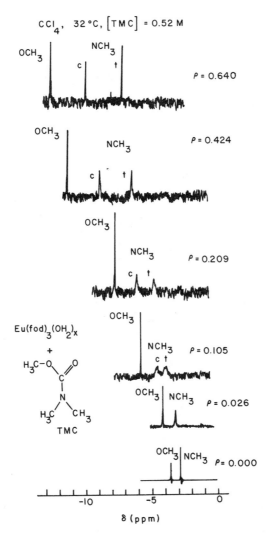

Figure 3. The proton NMR spectrum of TMC in CCℓ$_4$ with varying mole ratios (ρ) of Eu(fod)$_3$(OH$_2$)$_x$ (x < 1/2). [TMC] = 0.52 ± 0.02 M, T = 32°C. The spectrum was recorded with differing values of the spectrum amplitude setting for different values of ρ.

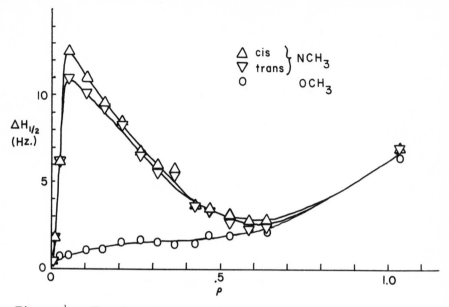

Figure 4.   The dependence of the full width at half-height
($\Delta H_{1/2}$) of the proton resonances of TMC on ρ.

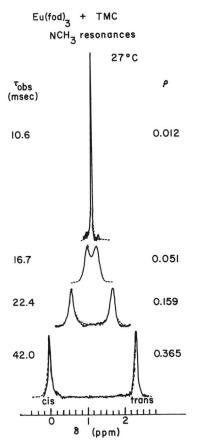

Figure 5. Experimental (——) and calculated (---) spectra of the N-CH$_3$ resonances in TMC with differing values of ρ. The values of τ given are those corresponding to the computer fit.

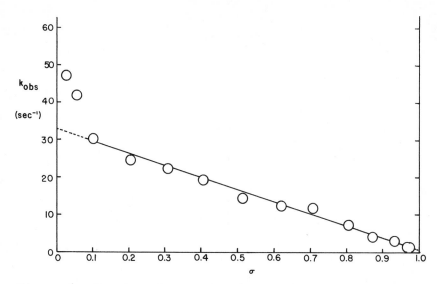

Figure 6. The dependence of the observed rate constant for isomerization of TMC, k_{obs}, upon the saturation fraction, the fraction of TMC coordinated to $Eu(fod)_3$, σ.

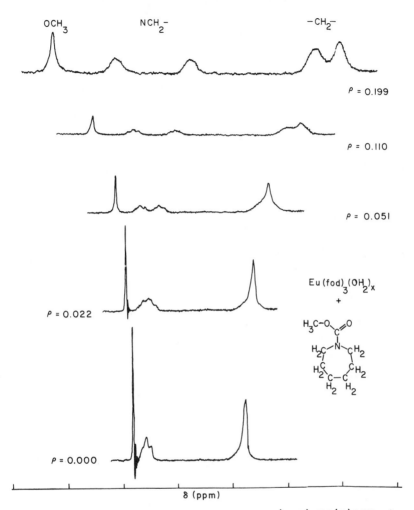

Figure 7. The proton NMR spectrum of $(H_2C)_6NC(O)OCH_3$ in CCl_4 with varying mole ratios (ρ) of $Eu(fod)_3$, T = 32°C.

311

DETERMINATION OF MOLECULAR CONFIGURATION FROM LANTHANIDE-INDUCED PROTON NMR CHEMICAL SHIFTS

Ian M. Armitage,[1] Laurance D. Hall,[2]
Alan G. Marshall,[*] and Lawrence G. Werbelow[3]

Department of Chemistry
University of British Columbia
Vancouver 8, Canada

INTRODUCTION

The determination of molecular configuration from lanthanide-induced proton NMR chemical shifts is severely underdetermined in at least two respects: first, the observed quantities represent average chemical shifts for a system of many coupled equilibria; second, the configuration of the molecule of interest is itself usually an average of several conformers. For nuclei other than ^{1}H, the third problem of substantial contributions from "contact" shift makes quantitative geometry determination virtually intractable. In this paper, we present what we believe to be the minimal chemical and mathematical compromises which render optimal geometrical information from a compact set of experimental measurements. Finally, as with any prescription, one must be alert to complications which may result when the treatment is applied to the wrong disease--in the final section, we list a set of practical criteria which should enable the user to recognize, in advance, systems which are likely to produce spurious or indeterminate answers.

PART I. DETERMINATION OF "BOUND" CHEMICAL SHIFTS FROM
         EXPERIMENTALLY MEASURED CHEMICAL SHIFTS.

"Bound" chemical shifts for each proton of the molecule of interest are the data which form the basis for subsequent geometry determinations, so their correct derivation is extremely important. The interaction between a lanthanide shift reagent, L, and a donor substrate, S, generally involves the following equilibria:

$$L + S \rightleftharpoons LS$$
$$LS + S \rightleftharpoons LS_2 \qquad\qquad [1]$$
$$L + L \rightleftharpoons L_2 \; .$$

Since all these processes are nearly always rapid compared to a typical "bound" shift (<u>ca.</u> 1000 Hz), the NMR "time-scale" dictates that the observed chemical shift for a given substrate proton must represent a concentration-weighted average of the chemical shifts of the respective species S, LS, and LS_2,

$$\delta_{obs} = f_S \Delta_S + f_{LS} \Delta_{LS} + f_{LS_2} \Delta_{LS_2} , \qquad [2]$$

where Δ represents the chemical shift of a substrate proton at the site of interest, and f is the fraction of substrate molecules at that site; the dimerization step is included in order to obtain the correct concentration fractions at each site. Detailed analysis of the scheme, [1], is given by J. Reuben elsewhere in this series. For present purposes, it suffices to note that extraction of the "bound" shifts, Δ_{LS} and Δ_{LS_2}, and the various equilibrium con-

stants from measurements of δ_{obs} requires solution of a
fifth-power equation, and thus an impractically large number of data points spanning a wide concentration range are
needed for valid interpretation.[4,5]

However, when (as in the present instance) the ulti-
mate object is molecular configuration, the experiment
should be re-designed to yield the "bound" shift, while
suppressing the complications from intermediate steps in
the formation of a shift reagent:substrate complex. The
simple condition,

$$[S]_0 \gg [L]_0 \qquad\qquad [3]$$

constrains the binding of substrate to shift reagent to
behave as a one-step process, with a single effective
binding constant and single bound shift:

$$L + n\ S \rightleftharpoons LS_n \quad, \quad n = 1\ \text{or}\ 2. \qquad [4]$$

From this condition, there are basically two ways to do
the experiment.

Method 1. Vary $[L]_0$ at constant $[S]_0$.

At least 99% of the reported "bound" shifts in the
literature are based on this type of experiment, from
which the "bound" shift is taken as the slope of a plot[6]
of δ versus $[L]_0/[S]_0$, where δ is the induced shift:

$$\delta = \delta_{obs} - \delta_S . \qquad\qquad [5]$$

Figure 1 illustrates two defects of this method. First of
all, if any impurity (typically water) is present in the
substrate solution, the impurity may bind shift reagent in
preference to the desired substrate, with the effect that
the plot is at best shifted toward the right on the x-axis,

and may be distorted as well, if the impurity has a binding constant comparable to that of the substrate. Second, the slope of such a plot (and thus the apparent "bound" shift) depends on the <u>absolute</u> substrate concentration in the experiment.[7] These effects are dramatically evident in the data[8,9] for Table 1, where it may be noted that the experiment carried out at the higher <u>absolute</u> substrate concentration yielded the larger apparent "bound" shifts-- more serious, the "bound" shift <u>ratios</u> from the two experiments differ markedly and lead to rather different apparent molecular geometries for this substrate. Finally, it is important to recognize that the most common present shift reagent,[10] $Eu(FOD)_3$, dimerizes extensively at the high shift reagent concentrations required in such experiments.[11] One is therefore faced with two unattractive alternatives inherent in this method of data reduction: either one uses $Eu(DPM)_3$, which does not dimerize,[11] but which has such low solubility that it becomes difficult to satisfy inequality [3] and still have high enough substrate concentration to obtain good signal-to-noise, or one resorts to the more soluble $Eu(FOD)_3$, and one is then beset by the problem of shift reagent dimerization; in either case, the presence of impurities will have a deleterious effect on the reliability of the apparent "bound" shift.

<u>Method 2.</u> Vary $[S]_0$ at constant $[L]_0$.

In contrast, by simply doing the experiment the other way around, one can neatly sidestep the problems of the previous method: by choosing a low (and constant) shift reagent concentration, dimerization of shift reagent may be neglected so that $Eu(FOD)_3$ may be used with impunity.

316

In addition, for the same concentration limit, $[S]_0 \gg [L]_0$, the algebra reduces to a simple linear equation,

$$[S]_0 = [L]_0 \Delta_i (1/\delta_i) - [(1/K_B) + [L]_0] \quad , \qquad [6]$$

so that a plot of $[S]_0$ versus $(1/\delta_i)$ for each substrate proton will have slope $[L]_0 \Delta_i$ and intercept, $[(1/K_B) + [L]_0]$, where K_B is the effective binding constant for the one-step binding process, [4]. A key feature of such a plot is that its slope (and thus the determined "bound" shift) is relatively insensitive to the presence of the impurities which vitiated the "bound" shift information in Method 1 (see J. Reuben, this series). Equation [6] and Method 2 thus afford the most reliable source of "bound" shifts from a compact set of experimental data--five data points suffice to give bound shifts to within better than 5%. The absolute error in these bound shifts is $(1 - (\delta_i / \Delta_i))$, which may be made arbitrarily small by choosing sufficient excess of substrate over shift reagent, and the error in shift ratios (the form in which the data are used in geometry calculations) will be even smaller.

The mathematical consistency of Method 2 is amply demonstrated by Figure 2. It may be noted that although the three substrate protons have markedly different bound shifts (and thus different slopes), the three lines intersect in a point, and that point lies on the y-axis, in exact accord with Eq. [6]. From this data reduction, we were able to show that the larger induced shifts typically observed using $Eu(FOD)_3$ compared to $Eu(DPM)_3$ were due to stronger binding of a given substrate to $Eu(FOD)_3$, since the bound shift was about the same in both cases.[12]

The <u>chemical</u> consistency of Method 2 is illustrated by Figure 3: although shift reagents containing different lanthanides give different bound shifts toward the same substrate (the shifts even differ in sign), the binding constant in each case is just about the same, in accord with the similarity in chemical properties across the lanthanide series.

It seems clear, then, that future attempts to obtain quantitative geometric information from lanthanide-induced "bound" chemical shifts should be based on experiments of the "Method 2" type--all "bound" shift data used in Part 2 of this paper were derived from plots of $[S]_0$ versus $(1/\delta_i)$ at constant $[L]_0$, as described at length in ref. 12.

****

PART II. DETERMINATION OF SUBSTRATE CONFIGURATION FROM "BOUND" CHEMICAL SHIFTS.

A number of explicit assumptions are basic to the calculation of geometry for a substrate:shift reagent complex.

(1) The "bound" shifts themselves are obtained in a reliable way (see Part I of this paper). Apart from the data reduction itself, correct results require some experimental precautions. We have found it necessary to prepare our own shift reagent from europium oxide, according to either ref. 13 [for $Eu(DPM)_3$] or ref. 14 [for $Eu(FOD)_3$]. In either case, the shift reagent for a particular experiment should always be sublimed immediately before use. All sample preparations are conducted in a glove bag which has

been flushed several times with dry nitrogen, and in which
a nitrogen atmosphere is maintained during preparations.
It is necessary to be scrupulous toward exclusion of
water--when possible, substrates should be distilled, sublimed, or recrystallized just before use and sealed or
kept in a vacuum desiccator over P_2O_5.

(2) The "bound" shift is wholly pseudo-contact ("dipolar")
in origin. This assumption is certainly invalid for
nuclei other than protons,[15] but appears to be generally
valid for proton shifts on two independent grounds. First,
observed proton shifts correlate nicely with dipolar shifts
calculated from single-crystal magnetic anisotropy data;[16]
second, since the pseudo-contact shift for Gd should be
near zero,[17] and since the contact shift for Gd is expected
to be nearly the largest for all the lanthanides,[18] any
observed shift using $Gd(DPM)_3$ should provide an upper limit
to the contact shift possible from $Eu(DPM)_3$. We have examined a number of picolines and lutidines (methyl- or dimethyl pyridines) in the presence of $Gd(DPM)_3$ and find
that the induced shifts are very small[19] and thus should
be negligible for $Eu(DPM)_3$, particularly for non-aromatic
substrates.

(3) The magnetic field in the substrate:shift reagent complex is axially symmetric, with principal axis along the
lanthanide-donor atom bond. Now it is well-known that
most solid-state X-ray and magnetic susceptibility studies
do not indicate axial symmetry,[20,21] but in solution, rapid
rotation about the lanthanide-donor bond need only be fast
compared to Δ_i (about 1000 Hz) in order that the <u>effective</u>
field at the nucleus be axially symmetric. This is readily

apparent from the explicit form for the "bound" shift for
an anisotropic molecular magnetic field:[22]

$$\Delta_i = A[(3\cos^2\theta_i - 1)/r_i^3] + B[\sin^2\theta_i \cos 2\psi_i/r_i^3]. \quad [7]$$

For fast internal rotation about the lanthanide-donor atom
bond, averaging $\cos 2\psi_i$ over 2π makes the second term of
Eq. [7] go to zero, leaving just the first term as the basis
for geometry calculations. In fact, the motion need only require rapid jumps between positions about a three-fold symmetry axis,[23] in order to make the second term of Eq. [7]
vanish. Accumulating X-ray evidence (R. E. Sievers et al,
this series), in the form of long lanthanide-donor bond
distances and also multiple forms of substrate:shift reagent crystals, points to substantial flexibility at the
site of attachment of substrate to shift reagent, with the
likelihood that rotational motion at that site may often
be fast enough to justify use of just the first term of
Eq. [7]. We have found no cases in which it has been
necessary to consider the second term of Eq. [7] in order
to obtain chemically reasonable structures for a substrate:
shift reagent complex.

With the above assumptions, Figure 4 defines the
starting point for determination of molecular geometry.
This right-handed coordinate system has been designed to
facilitate computer fits of "bound" shift data: the donor
atom (atom 1) defines the origin; proceeding from atom 1 to
atom 2 defines the positive x-direction; atom 3 is then as-
signed a positive y-value in the x-y plane. Ω, ϕ, and $\underline{R}$
unambiguously fix the position of the lanthanide atom rela-
tive to the molecular frame.

Three substrate molecules will suffice to illustrate
the principles involved in geometry determinations--the
results will indicate an optimal line of attack for substrates of unknown configuration where various rotamers
or conformers are present.

The illustrative molecules are chosen to be rigid
everywhere except at the point of attachment to the lanthanide, and a "determination" of the configuration of the
complex thus consists of finding the "best" values of R, Ω,
and ϕ, given the structure of the substrate itself. Ref.
29 gives a rapid method for obtaining the desired parameters r_i and θ_i for the i'th proton, from (guessed) values
of the Eu-donor atom bond distance R, the "flex" (Ω) and
"twist" (ϕ) angles which locate the Eu-donor bond axis with
respect to the molecular frame, and the coordinates of all
atoms in the substrate. If the complex were rigid, one
could proceed as follows. First, guess a location for the
lanthanide atom (i.e., choose values for R, Ω, and ϕ) and
compute $[(3\cos^2\theta_i - 1)/r_i^3]$ for each proton of the substrate; then calculate the normalized variance (the "R"-
value)[24] between ratios of this quantity and observed shift
ratios for all possible independent pairs of protons. Then
repeat the procedure many times for different values of R,
Ω, and ϕ, and choose the most probable single conformation
as that which gives the best "fit" (smallest normalized
variance, smallest "R"-value) to the observed shift ratios.
The difficulty with this treatment is that it is quite possible to obtain correct shift ratios from incorrect
absolute shifts, so that sometimes the best "fits" are obtained at chemically unreasonable values of R and Ω.

The source of the difficulty lies in attempting to fit the observed shift ratios to those computed for <u>individual</u> conformations (this procedure will at best provide an average value for r_i and θ_i). However, the observed quantity is Δ_i, where

$$\Delta_i \; \alpha \; \left\langle \frac{3\cos^2\theta_i - 1}{r_i{}^3} \right\rangle \neq \frac{3\cos^2\langle\theta_i\rangle - 1}{\langle r_i\rangle^3} \, , \qquad\qquad [8]$$

where the brackets denote an average over all possible "bound" conformations during the residence of a substrate at a shift reagent site. Whenever rapid internal rotations are present, it is necessary to average the <u>entire quantity</u> $[(3\cos^2\theta_i - 1)/r_i{}^3]$ <u>before</u> comparing observed and calculated shift ratios; because of inequality [8], any analysis based on a best single conformation is not expected to succeed. The simplest models for internal rotation are: no rotation at all, free rotation over 2π range of ϕ, rotation over a limited range of ϕ, and rotation restricted to jumps between minima of an n-fold potential. We will now show how to resolve a molecular problem in terms of these models.

Free rotation about the donor-carbon axis (atom 1 to atom 2 in Figure 4) is readily simulated by multiplying the quantity, $[(3\cos^2\theta_i - 1)/r_i{}^3]$ by a normalized unit weight factor,

$$P(\phi) \; d\phi = (1/2\pi) \; d\phi \, , \qquad\qquad [9]$$

followed by numerical integration over all ϕ from 0 to 2π, where the integration is carried out <u>before</u> comparing observed and calculated shift ratios. The ordinary method for the other extreme limit of a rigidly locked complex would correspond to use of a weight factor,

322

$$P(\phi) \, d\phi = \delta(\phi - \phi_0) \, d\phi \, , \qquad\qquad [10]$$

where ϕ_0 is the fixed "twist" angle in the Dirac delta function. Since no real molecule will be perfectly rigid, it is desirable to relax the distribution, [10], to span some specified angular range in the vicinity of ϕ_0--we have chosen, for convenience, a Gaussian weight factor:[25]

$$P(\phi)d\phi = (A/\sqrt{\pi}) \, \exp[-A^2(\phi-\phi_0)^2] \, d\phi \, . \qquad [11]$$

In Eq. [11], a large value of A corresponds to a narrow distribution of angular positions; the A-values of $8^{1/2}$ or 1 in the next section correspond to rms widths of about 14° or 40° about ϕ_0, respectively. Finally, random rapid jumps between various ϕ-values are most simply simulated by

$$P(\phi)d\phi = a\delta(\phi-\phi_1) + b\delta(\phi-\phi_2) +\cdots+ f\delta(\phi-\phi_n)d\phi, \qquad [12]$$

where a, b, $\cdots$, f represent the probability of finding a "twist" angle of ϕ_1, ϕ_2, $\cdots$, or ϕ_n, respectively. Numerical integration (when necessary) is efficiently achieved using low-order Gauss-Legendre quadrature (i.e., 6, 8, or 10 point), is checked against higher order formulas to verify validity, and costs less than $1.00 U.S. in IBM 360-67 computer time. COMPLETE COMPUTER PROGRAMS FOR THESE PROCEDURES ARE AVAILABLE ON REQUEST!

****

PART III. RESULTS AND DISCUSSION.

Determinations of molecular geometry from chemical shift ratios are best evaluated from contour plots of the type shown in Figures 5-8. The contours are simply paths of constant normalized variance ("R"-value, agreement factor) between observed and calculated shift ratios, as a function of possible positions of the lanthanide-donor atom distance ($\underline{R}$), the "flex" angle (Ω) between the Eu:donor bond and the carbon-donor bond, and the "twist" angle (ϕ) shown in Figure 4. A "valley" in the contour map corresponds to best agreement between observed and calculated shift ratios.

The first substrate considered is the monofunctional donor, diacetoneglucose, or DAG.[10] Figure 5 shows the

DAG

contours which are obtained, under the assumption that there is no internal rotation about the carbon-donor bond. Two features are evident. First, for some choices of ϕ, there are simply no "good" fits (i.e., having a normalized variance of 0.04 or less). Second, among the range of ϕ-values for which good fits are obtained, some ϕ-values lead to unreasonably short europium-oxygen bond distances.[21] Based on Figure 5, if DAG is rigid with respect to internal rotation about the carbon-donor bond, then the most likely position for europium is $\underline{R}$ = 2.2 Å, Ω = 114°, and ϕ = 116°.

Figure 6 shows the effect of varying degrees of internal motion on the agreement between observed and calculated shift ratios for DAG. Beginning (as in Figure 5) with a static molecular frame, we now allow for a Gaussian distribution of ϕ-values, centered at the most likely ϕ-value of 116°, with a root-mean-square width of either 14°("narrow"

Gaussian in Figure 6) or 40° ("wide" Gaussian). It is
clear that this greater latitude in internal rotational
position produces less reasonable fits, with respect both
to agreement with experiment (large normalized variance)
and also intuition (too-short Eu-O bond distance).[21] In
fact, the contour plot for the assumption of completely
free internal rotation about the carbon-donor bond (bottom
right in Fig. 6) shows that free rotation is simply not
possible in this adduct. Figures 5 and 6, then, demonstrate conclusively that the DAG:Eu(DPM)$_3$ complex is relatively rigid, so that a unique geometry for the adduct may
be determined with confidence.

For the binding of aniline to Eu(DPM)$_3$, the situation
is quite reversed. In this case, an assumption of completely free internal rotation about the C-N bond yields a
very good "fit", as seen in the top plot of Figure 7. The
resultant value of 2.55 Å for the Eu-N bond distance com-
pares favorably with typical X-ray values of about 2.65Å.[21]
Now since the experiment shows a single resonance for protons 2 and 6 (or 3 and 5) on opposite sides of the aromatic
ring, the only possible unique position for Eu in a rigid
complex would be for ϕ = 90°; however, the middle plot of
Fig. 7 shows that this idea is simply not tenable in view
of the poor agreement with experimental shift ratios. The
remaining possibility is that of random jumps between (for
example) ϕ-values of 0 and 180°; contours for this last
model appear as the bottom plot of Fig. 7. The "fit" for
this "jump" model is very sharply defined, with $\underline{R}$ = 2.75 Å
and Ω = 115°. In conclusion for aniline, no static conformation will fit the experimental data, and models of

either free rotation or random jumps between ϕ = 0 and 180° produce comparable (good) results which bracket the X-ray value for the Eu-N bond distance.

Figure 8 illustrates a final situation of interest. The top two plots indicate that good "fits" are obtained from either a "static" or free rotation model for motion about the carbon-donor oxygen bond for the adduct formed between compound 1 and Eu(FOD)$_3$. However, the very best

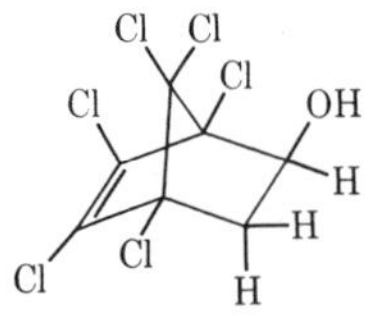

1

"fits" (smallest normalized variance) were obtained in the bottom plot, which is a Gaussian distribution in ϕ centered at ϕ = 248° with rms width of 40 °. The embarrassing feature of this plot is the very wide range in values of R and Ω over which equally good fits can be obtained, so that virtually no information about R or Ω or internal conformation can be derived from this particular experiment. For substrate 1, the bound conformation probably exhibits some internal rotation, with the Eu more often opposed than adjacent to the apical chlorines.

****

CONCLUSIONS

A number of procedural conclusions may be drawn from the experiments described in this paper:
(1) For reliable results, the "bound" shifts should be determined from a plot of $[S]_o$ versus $(1/\delta)$ for experiments

in which $[L]_0$ is kept small and constant, while $[S]_0$ is varied.[12] It may be noted that this requirement would exclude virtually all existing data in the literature. For polyfunctional molecules, $Eu(DPM)_3$ may be preferred to $Eu(FOD)_3$. While some attention has been given to such cases,[26-28] it appears that the problem will be too complex unless (a) the two binding sites are spatially far apart to avoid interactions between binding sites, and (b) the two functional groups have binding constants which are sufficiently different (say a factor of 10) that it, is possible to bind shift reagent selectively to one site only.

(2) Contact shifts for protons may generally be ignored: they appear negligible for $Eu(DPM)_3$ even with substituted pyridine substrates.[16,20] It is thus not necessary to perform measurements with shift reagents which produce shifts of opposite sign in order to obtain reliable results. Contact shifts for other nuclei are serious, and probably will preclude the quantitative use of ^{13}C, ^{19}F, or ^{31}P induced shifts in geometry determination,[15,27] unless other shift reagents such as $Pr(FOD)_3$ or $Yb(FOD)_3$ are employed.[30] Diamagnetic shifts (as might be seen from $La(DPM)_3$) have never been observed and are almost certainly negligible.

(3) Magnetic symmetry in the substrate:shift reagent adduct, while probably not axially symmetric at a given instant, is expected in most cases to average to effectively axial symmetry due to motion about the lanthanide-donor bond (with resultant principal magnetic axis along the

lanthanide-donor bond)with the possible exception of substrates which exhibit serious steric hindrance at the site
of binding to shift reagent. However, even for the present
example of DAG, which binds rigidly to $Eu(DPM)_3$, successful
data analysis could be achieved by use of a one-term
equation for the dipolar shift; the first term of Eq. [7]
will probably suffice for the great majority of substrates.
(4) Most important, the present results[29] show that since
it is generally possible to "fit" the experimental shift
ratios using the known X-ray value (Chap. 2) for a Eu-O or
Eu-N distance and the now-established "flex" angle between
either Eu-O and C-O or Eu-N and C-N for a given functional
group, a whole new spectrum of applications brightens the
future use of lanthanide NMR shift reagents. From now on,
we suggest that the following procedure may be applied
toward determination of the solution geometry of molecules
whose conformation is not known in advance! Using the
X-ray value for (say) the Eu-N distance and a suitable
"flex" angle for the substrate of interest, one can then
construct contour plots of the type shown in Figures 5-8,
based on Eqs. [9-12], where the new independent variables
now represent various internal rotamers or ring conformers,
and where proton-induced shifts form the basis for the analysis. Consequent computer-based comparisons between observed and calculated "bound" shift ratios, contrary to
general belief, are cheap and relatively straightforward,
and copies of our own programs are available on request.

ACKNOWLEDGMENTS: This work was supported by grants (to
AGM and LDH) from the National Research Council of Canada
and the University of British Columbia.

328

FOOTNOTES AND REFERENCES

1. Present address: Department of Chemistry, California Institute of Technology, Pasadena, California 91030.

2. Alfred P. Sloan Foundation Research Fellow, 1971-73.

3. University of British Columbia Graduate Fellow, 1972-3.

4. D. A. Deranleau, J. Amer. Chem. Soc. 91, 4044 (1969).

5. B. L. Shapiro and M. D. Johnston, Jr., J. Amer. Chem. Soc. 94, 8185 (1972).

6. P. V. Demarco, T. K. Elzey, R. B. Lewis, and E. Wenkert, J. Amer. Chem. Soc. 92, 5734, 5737 (1970).

7. I. M. Armitage, G. Dunsmore, L. D. Hall and A. G. Marshall, Chem. Commun. 1281 (1971).

8. J. Reuben and J. S. Leigh, Jr., J. Amer. Chem. Soc. 94, 2789 (1972).

9. R. A. Fletton, G. F. H. Green, and J. E. Page, Chem. Ind. (London), 167 (1972).

10. Abbreviations used: Eu(DPM)$_3$ [also known in the literature as Eu(thd)$_3$] for tris(dipivalomethanato)-europium (III), also as tris(2,2,6,6-tetramethyl-3,5-heptanedionato)Eu(III); Eu(FOD)$_3$ for tris(6,6,7,7,8,8,8-heptafluoro-2,2-dimethyl-3,5-octanedionato)europium(III); DAG for diacetoneglucose, proper name 1,2:5,6-di-O-isopropylidene-α-D-glucofuranose; 1 for exo-5-hydroxy-1,2,3,4,7,7-hexachloronorborn-2-ene.

11. J. F. Desreux, L. E. Fox, and C. N. Reilley, Analyt. Chem. 44, 2217 (1972).

12. I. M. Armitage, G. Dunsmore, L. D. Hall, and A. G. Marshall, Can. J. Chem. 50, 2119 (1972).

13. K. J. Eisentraut and R. E. Sievers, J. Amer. Chem. Soc. 87, 5254 (1965).

14. L. S. Springer, Jr., D. W. Meek, and R. E. Sievers, Inorg. Chem. 6, 1105 (1967).

15. G. E. Hawkes, C. Marzin, S. R. Johns, and J. D. Roberts, J. Amer. Chem. Soc. 95, 1661 (1973).

16. W. DeW. Horrocks, Jr., and J. P. Sipe, III, Science 177, 994 (1972).

17. W. DeW. Horrocks, Jr., and J. P. Sipe, III, J. Amer. Chem. Soc. 93, 6800 (1971).

18. R. M. Golding and M. P. Halton, Austral. J. Chem. 25, 2577 (1972).

19. A. G. Marshall and S. K. Foster, unpublished results.

20. W. DeW. Horrocks, Jr., J. P. Sipe, III, and J. R. Luber, J. Amer. Chem. Soc. 93, 5258 (1971).

21. R. E. Cramer and K. Seff, Chem. Commun. 400 (1972).

22. W. DeW. Horrocks, Jr., and E. S. Greenberg, Inorg. Chem. 10, 2190 (1971).

23. J. M. Briggs, G. P. Moss, E. W. Randall, and K. D. Sales, Chem. Commun. 1180 (1972).

24. M. R. Willcott, III, R. E. Lenkinski, and R. E. Davis, J. Amer. Chem. Soc. 94, 1742 (1972); R. E. Davis and M. R. Willcott, III, ibid. 94, 1744 (1972).

25. The actual domain of integration is from $\phi = \phi_0 + \pi$ to $\phi = \phi_0 - \pi$. Thus it is correct to compute induced shift ratios for any given A, but one should not compare absolute shifts computed for different choices of A.

26. C. C. Hinckley, M. R. Klotz, and F. Patil, J. Amer. Chem. Soc. 93, 2417 (1971).

27. J. K. M. Sanders, S. W. Hanson, and D. H. Williams, J. Amer. Chem. Soc. 94, 5325 (1972).

28. J. K. M. Sanders and D. H. Williams, Nature 240, 385 (1972).

29. I. M. Armitage, L. D. Hall, A. G. Marshall, and L. G. Werbelow, J. Amer. Chem. Soc. 95, 1437 (1973).

30. G. E. Hawkes, D. L. Leibfritz, D. W. Roberts, and J. D. Roberts, J. Amer. Chem. Soc. 95, 1659 (1973). See also Chapter 6.

Table 1. Eu(DPM)$_3$ - Induced Proton NMR Chemical Shifts
(ppm) for pyridine substrate. [See Part I. of text].

Proton			Induced Shift Ratios		$[S]_0$	Reference
o	m	p	o/m	o/p		
25.9	9.0	8.2	2.88	3.16	0.2 M	8
28.7	9.9	9.2	2.90	3.12	0.5 M	9

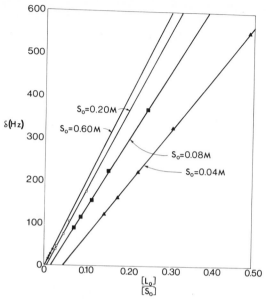

Figure 1. Plot of induced chemical shift (δ) versus ratio of the initial concentrations of Eu(DPM)$_3$, [L]$_0$, to <u>neo</u>-pentanol, [S]$_0$, in deuteriochloroform solvent. Points on a given line represent experiments in which the concentration of shift reagent was varied, keeping the neopentanol concentration fixed at the value listed for that line: see Part I, Method 1 of text.

332

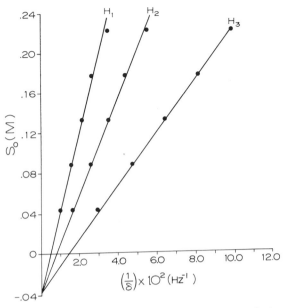

Figure 2. Plot of [S]$_0$ versus (1/δ) for the interaction of n-propylamine (0.23 to 0.04 M) with Eu(DPM)$_3$ (ca. 0.006 M) in deuteriochloroform solvent. The separate lines are for protons H$_1$, H$_2$, and H$_3$ of the substrate. Data reduction corresponds to Method 2, in Part I. of the text.

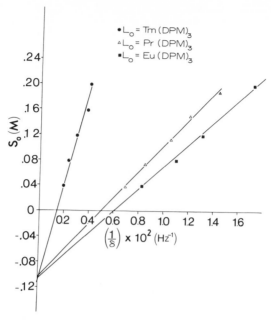

Figure 3. Plot of [S]$_0$ versus (1/δ) for the interaction of neo-pentanol with three different lanthanide shift reagents (each ca. 0.006 M). For Eu(DPM)$_3$ (downfield shift) and Tm(DPM)$_3$ (upfield shift), CDCl$_3$ was used as solvent, while for Pr(DPM)$_3$ (upfield shift), CHCl$_3$ was used as solvent. All shifts are for the H$_1$ proton of neo-pentanol.

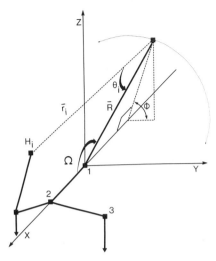

Figure 4. Coordinate system for substrate:shift reagent complex. Lanthanide atom is at upper right; atom 1 is the donor atom (usually N or O). $\underline{R}$ is the lanthanide-donor bond vector, $\bar{r}_i$ is the lanthanide-proton distance vector, and θ_i is the angle between $\underline{R}$ and $\bar{r}_i$. Ω is the "flex" angle between $\underline{R}$ and the donor-carbon bond, and internal rotation of $\underline{R}$ about the x-axis (shown as circular arc) consists of allowing a range of "twist" angle (ϕ) positions. Vector from atom 1 to atom 2 defines the positive x-direction; atom 3 is then assigned a positive y-value in the x-y plane; z-direction then follows from a right-hand convention.

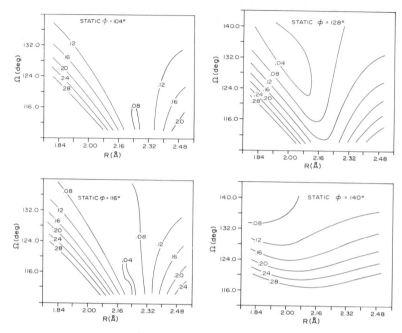

<u>Figure 5.</u> Contours of normalized variance (between ob-
served and calculated "bound" shift ratios) as a function
of possible positions of the lanthanide atom relative to
the donor atom of the substrate, DAG.[10] It has been assumed that there is no internal rotation about the bond
from carbon to donor oxygen. R, Ω, and φ are as in Fig. 4.

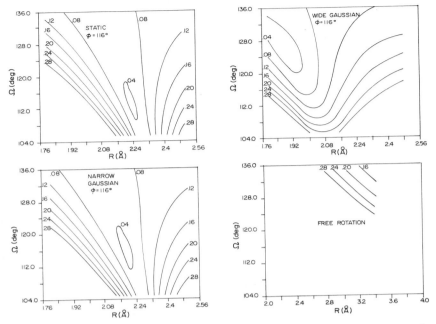

<u>Figure 6.</u> Contours of normalized variance as a function of lanthanide position for DAG.[10] For the "static" plot, there is no internal rotation about the carbon-donor bond. For the "narrow" and "wide" Gaussian, ϕ-values are first weighted by the factor, $(A/\sqrt{\pi})\exp[-A^2(\phi-\phi_0)^2]d\phi$ and the corresponding $[(3\cos^2\theta_i - 1)/r_i^3]$ is then integrated over all ϕ (with $A = \sqrt{8}$ or 1, respectively) before comparing observed and calculated shift ratios. For the "free rotation" plot, the dipolar shift is averaged over all ϕ from 0 to 2π using unit weight factor. (See text, Part II.)

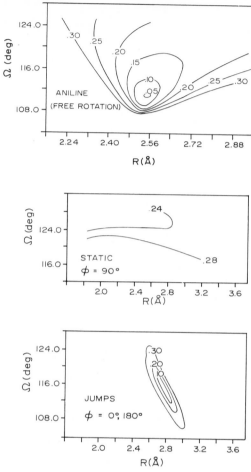

<u>Figure 7.</u> Contours of normalized variance as a function of lanthanide position for aniline with Eu(DPM)$_3$. Top: internal rotation about the C-N bond is assumed to be completely free (unhindered). Middle: Eu is assigned a single ϕ-value of 90° (rigid complex). Bottom: Eu is assumed to jump rapidly between $\phi = 0°$ and $\phi = 180°$.

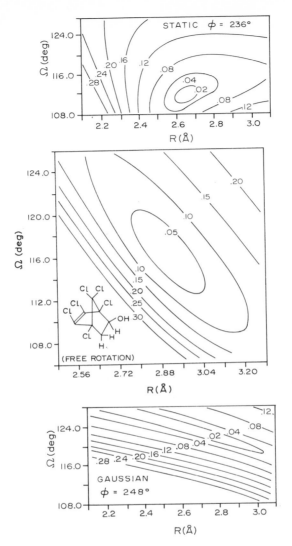

Figure 8. Contours of normalized variance as a function of lanthanide position for binding of $\underline{1}^{10}$ to Eu(FOD)$_3$.
Top: no internal rotation about the C-O bond.
Middle: free rotation about the C-O bond.
Bottom: Gaussian distribution in "twist" position of Eu about the C-O bond, centered at $\phi = 248°$ with rms width of about 40°.

EFFECTS OF CHEMICAL EQUILIBRIUM AND ADDUCT

STOICHIOMETRY IN SHIFT REAGENT STUDIES

Jacques Reuben

Isotope Department, The Weizmann Institute of Science,
Rehovot, Israel

Introduction

Paramagnetic lanthanide β-diketonates of the
Ln(dpm)₃ and Ln(fod)₃ types have found a wide variety of
applications as NMR shift reagents [1]. Their usefulness
depends both on the magnetic properties of the central ion
and on the chemical properties of the complex as a whole,
which in turn reflect the nature of the central ion and of
the surrounding ligands. The action of these complexes as
shift reagents depends entirely upon their ability to form
labile adducts with organic molecules (substrates) in
solution. Thus the shifts induced in the NMR spectrum of
the substrate molecule are dependent to a large extent on
the adduct stoichiometry and on the chemical equilibria
present in the observed system. Recent work has shown that
Eu(fod)₃ and Pr(fod)₃ form 2:1 substrate-reagent adducts in
solution [2-7]. Adducts of 2:1 stoichiometry in the solid
state have been obtained with Ln(dpm)₃, however, their
existence in solution is not firmly established [1]. It is
definitely preferable to use shift reagents forming only
1:1 adducts of relatively low dissociation constant. It is
also desirable that the induced shifts be of dipolar origin
since dipolar shifts are directly related to molecular geometry. Although Eu(dpm)₃ and Eu(fod)₃ are the most widely
used compounds they are by no means ideal and the search
for the "best" shift reagent is continuing.

In this paper we discuss the problem of the
effects of chemical equilibrium and adduct stoichiometry in
studies of shift reagents. The full equations describing

these effects upon observable quantities are used to simulate experimental conditions (for a series of simple cases) thereby allowing a critical examination of approximate procedures recommended in the literature. Contrary to the notion [8] that on the average only 1:1 adducts can be detected in systems undergoing rapid chemical exchange, it is shown that the more complex situation of the formation of 2:1 adducts is also observable.

Equilibria in Shift Reagent Systems

The average position of an NMR spectral line due to nuclei rapidly exchanging between several distinct chemical environments is the average of the line positions corresponding to each of these environments weighted by their relative populations. For a nucleus of a substrate molecule in the presence of a shift reagent forming 2:1 adducts the chemical shift relative to that in the absence of reagent is given by

$$\Delta = [RS]\Delta_1/S_t + 2[RS_2]\Delta_2/S_t \quad , \tag{1}$$

where S_t is the total substrate concentration and $[RS]$, Δ_1 and $[RS_2]$, Δ_2 are respectively the equilibrium concentrations and limiting shifts of the 1:1 and 2:1 adducts.

First we consider the simple adduct formation equilibria

$$R + S \rightleftharpoons RS , \quad K_1 = [R][S]/[RS] , \tag{2}$$

$$RS + S \rightleftharpoons RS_2 , \quad K_2 = [RS][S]/[RS_2], \tag{3}$$

where the K's are dissociation constants. The simplest way to treat the system is to solve for $[S]$ obtaining

$$[S]^3 + (2R_t - S_t + K_2)[S]^2 + K_2(R_t - S_t + K_1)[S]$$

$$- K_1K_2 S_t = 0 , \tag{4}$$

where R_t is the total reagent concentration. With $[S]$ the other components of the system are easily obtained, e.g.

$$[R] = R_t/(1 + [S]/K_1 + [S]^2/K_1K_2) . \qquad (5)$$

If $K_2 = 4K_1$ then the reagent has two equivalent and independent sites for substrate coordination, characterized by an intrinsic (statistically corrected) adduct dissociation constant $K = K_1/2 = 2K_2$. It can be easily shown that for $\underline{n}$ such sites

$$S_t - [S] = [S_t + nR_t + K - \sqrt{(S_t + nR_t + K)^2 - 4nS_tR_t}]/2 . \qquad (6)$$

In addition to adduct formation there are at least three other types of chemical interactions that are often present in shift reagent - substrate systems. These include the presence of another substrate, e.g. water impurity, complex formation between the substrate and the complex, e.g. hydrogen bonding to chloroform, and dimerization of the reagent. We will consider the effects of such interactions for the case n = 1.

The presence of an impurity binding to the reagent will add the equilibrium

$$R + I \underset{\rightarrow}{\leftarrow} RI , \quad K_i = [R][I]/[RI] , \qquad (7)$$

which for n = 1 will exist simultaneously with that described by equation 2. Solving for [R] we obtain

$$[R]^3 + (K_1 + K_i + S_t + I_t - R_t)[R]^2$$

$$+ [K_1K_i + S_tK_i + I_tK_1 - R_t(K_1 + K_i)][R] - R_tK_1K_i = 0 . \qquad (8)$$

With [R] the concentrations of other components of the system can be calculated, e.g.,

$$[RS] = S_t[R]/(K_1 + [R]) . \qquad (9)$$

Equation 9 holds also for the case of reagent dimerization which adds (to equilibrium 2) the equilibrium

$$R + R \rightleftarrows R_2 , \quad K_d = [R]^2 / [R_2] . \tag{10}$$

With it the expression for $[R]$ is of the form

$$2[R]^3 + (2K_1 + K_d)[R]^2 + K_d(K_1 + S_t - R_t) [R]$$

$$- R_t K_1 K = 0 . \tag{11}$$

Note that it is assumed that the dimer does not coordinate substrate.

Interaction with solvent, A , adds the equilibrium

$$S + A \rightleftarrows SA , \quad K_s = [S][A]/[SA] . \tag{12}$$

Solving equations 2 and 12 for $[S]$ we obtain

$$[S]^3 + (K_1 + K_s + R_t + A_t - S_t) [S]^2$$

$$+ [K_1 K_s + R_t K_s + A_t K_1 - S_t (K_1 + K_s)][S] - S_t K_1 K_s = 0 . \tag{13}$$

The three interactions above were considered as existing all separately. If present simultaneously the equations become rather involved. The case of reagent dimerization coupled to the formation of 2:1 adducts of the general type ($K_2 \neq 4K_1$) has been treated [6].

Simulation and Discussion of Experiments

It has become customary in the art to plot induced shifts, Δ , measured at constant substrate concentration against the molar ratio of reagent to substrate. It can be shown that for the case of $K_2 = 4K_1 = 2K$ and $\Delta_1 = \Delta_2 \equiv \Delta_M$ the shift as a function of ρ is given by

$$\Delta = \frac{n \rho \Delta_M}{1 + K/(S_t - S_b)} , \tag{14}$$

where S_b = [RS] for n = 1 and S_b = [RS] + 2[RS$_2$] for n = 2.
At sufficiently high values of ρ $\Delta \approx \Delta_M$. At very low
values of ρ there will be an (almost) linear portion of
slope $n\Delta_M/(1 + K/S_t)$ and if the concentration is chosen
such that $K/S_t < < 1$ then from the slope and limiting
shift the value of n can be estimated. In this way it has
been found that $1 < n < 2$ for Eu(fod)$_3$ and Pr(fod)$_3$ [6,7].
Curve fitting is needed in order to obtain K and accurate
Δ_M. A simulation of typical (albeit simple) cases is shown
in Figure 1, where for the 2:1 adduct the separate contributions of the RS and RS$_2$ species are also given. There
are several noteworthy features. The effect of the
solvent-substrate interaction is dramatic. Indeed, smaller
induced shifts in otherwise identical conditions have been
observed (in one case going down by a factor of 2.2 [9])in
changing the solvent from CCl$_4$ to CDCl$_3$ [1]. If n = 2 but
$\Delta_1 \neq \Delta_2$ then curves similar to those shown for the RS or
RS$_2$ components may be observed, depending on whether
$\Delta_1 > \Delta_2$ or $\Delta_1 < \Delta_2$. In general different protons in the
same molecule may have different Δ_1/Δ_2 ratios as has been
found with quinoline [3,8]. Finally, the shape of the
curve obtained in the presence of an impurity-reagent
interaction is very similar (almost coinciding) to that of
the RS component for the n = 2 case and may lead to
erroneous interpretations of experimental results.

Equation 14 can be rearranged in the form

$$S_t(1 - \Delta/\Delta_M) = nR_t(\Delta_M/\Delta - 1) - K \quad . \qquad (15)$$

If experiments are carried out at constant R_t and varying
S_t such that $S_t > > R_t$, then $\Delta/\Delta_M < < 1$ ($\Delta_M/\Delta > > 1$) and
Equation 15 simplifies to

$$S_t = nR_t\Delta_M/\Delta - K \quad . \qquad (16)$$

Thus the plot of S_t against $1/\Delta$ will be linear with a slope
of $nR_t \Delta_M$ and an intercept of $-K$. This method has been
recommended by Armitage et al. [10]. A simulation is shown
in Figure 2. It is seen that at low values of S_t
a curvature becomes evident since the condition of
$\Delta/\Delta_M < < 1$ is not fulfilled. The effect of dimerization is

remarkably small since R_t is very low. The interactions
with impurity or solvent have little effect on the slope.
Thus this method may serve to obtain reliable values of Δ_M
but only within a factor of n, since there is no way of
separating the product $n\Delta_M$. It has been suggested that
curvature in S_t versus $1/\Delta$ plots is indicative of 2:1
stoichiometry [5, 10c]. This will be correct only if
$\Delta_1 \neq \Delta_2$. We have simulated on Figure 3 a case of $n = 2$
with $K = 40$ mM and given separately the RS and RS_2 contributions. It is clearly seen that each of them shows easily
observable curvature, which is virtually absent in the
total reciprocal shift. Thus if only the RS_2 species are
considered [10c] a curvilinear plot will result. Simulations show that these curvatures are more pronounced at
higher K's. The surprising change of stoichiometry with
changing solvent reported by Gibb et al. [5] can now be
rationalized. The solvent (chloroform) effect is to
reduce the effective K thereby revealing the curvature.
Thus the total stoichiometry in this case is also 2:1 as
found for Eu(fod)$_3$ and other substrates but probably
$\Delta_2 > \Delta_1$. The fact that the Δ_M values reported for CCl_4
solutions are about twice as large as those for $CDCl_3$ [5]
comes from the assumption that $n = 1$ in the former case and
$n = 2$ in the latter. As pointed out above the slope of S_t
versus $1/\Delta$ curves gives only the product $n\Delta_M$. Summarizing
this point: the presence of substantial curvature in the
half-reciprocal plots is indicative of 2:1 stoichiometry
with $\Delta_1 \neq \Delta_2$, its absence however leaves the problem
insufficiently determined.

A combination of Equations 1 and 6 for the case
$\Delta_1 = \Delta_2$ gives

$$\Delta = \Delta_M[1 + n\rho + \rho K/R_t - \sqrt{(1 + n\rho + \rho K/R_t)^2 - 4n\rho}]/2 . \quad (17)$$

If $n = 1$ and experiments are done such that $\rho = 1$ throughout
then equation 16 can be rearranged in the form [11]

$$\Delta = \Delta_M - \sqrt{\Delta_M K_1 (\Delta/R_t)} \quad . \quad (18)$$

Thus a plot of Δ against $\sqrt{\Delta/R_t}$ is linear with intercepts Δ_M
on the ordinate and $\sqrt{\Delta_M K_1}$ on the abscissa. If $n = 2$ and
$\rho = 1/2$ again a linear Δ versus $\sqrt{\Delta/R_t}$ relationship is
obtained:

$$\Delta = \Delta_M - \sqrt{(\Delta_M K/2)(\Delta/R_t)} \quad . \tag{19}$$

These are modified forms of the Scatchard equation [1,6]:

$$\frac{\Delta}{(\Delta_M - \Delta)R_t} = \frac{n}{K} - \frac{\Delta}{K\rho\Delta_M} \quad . \tag{20}$$

Two series of experiments done with $\rho = 1/2$ and with $\rho = 1$ should immediately point out the stoichiometry and give as well the limiting shift and the adduct dissociation constant. Simulations are shown in Figure 4. It is seen that Δ extrapolates to Δ_M for $n = 1$, when $\rho = 1$ and for $n = 2$ both when $\rho = 1/2$ and when $\rho = 1$, however, for $n = 1$ and $\rho = 1/2$, Δ approaches $\Delta_M/2$ as can be expected intuitively. It has been our experience that such experiments are meaningful only if the concentrations can be varied over a wide range encompassing sufficiently large portions of the saturation curve. In many cases this may be difficult since ρ is maintained constant and K is of the order of 10 mM. Mere linearity of the plot over a narrow range [11] is of course inconclusive.

In the above discussion we have considered some idealized cases. In practice, however, the more complex situations of $\Delta_1 \neq \Delta_2$ and $K_2 \neq 4K_1$ have been encountered. The analysis of such systems is cumbersome but nevertheless manageable [4,6].

References

1. For a recent review see: J. Reuben, Progr. NMR Spectr., 9, 1 (1973).

2. D.F. Evans and M. Wyatt, Chem. Commun., 312 (1972).

3. K. Roth, M. Grosse and D. Rewicki, Tetrahedron Letters, 435 (1972).

4. B.L. Shapiro and M.D. Johnston, Jr., J. Amer. Chem. Soc., 94, 8185 (1972).

5. V.G. Gibb, I.M. Armitage, L.D. Hall and A.G. Marshall, ibid., 94, 8919 (1972).

6. J. Reuben, ibid., 95, (1973), in press.

7. J.W. ApSimon, H. Beierbeck, and A. Fruchier, ibid., 95, 939 (1973).

8. J.K.M. Sanders, S.W. Hanson, and D.H. Williams, ibid., 94, 5325 (1972).

9. I. Armitage and L.D. Hall, Can. J. Chem., 49, 2770 (1971).

10. I. Armitage, G. Dunsmore, L.D. Hall and A.G. Marshall, (a) Chem. Commun., 1281 (1971); (b) Chem. Ind. 79 (1972); (c) Can. J. Chem., 50, 2119 (1972).

11. J. Bouquant and J. Chuche, Tetrahedron Letters, 2337 (1972).

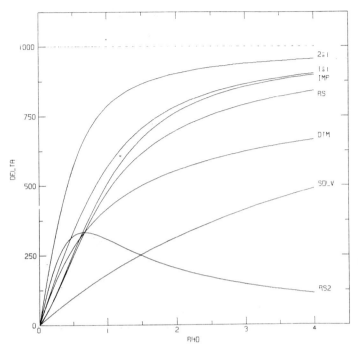

<u>Figure 1</u>   Calculated curves of Δ versus ρ . The
parameters used in the calculations are Δ_M =
1000 Hz, S_t = 30 mM, K = 10 mM, I_t = 5 mM,
A_t = 10 mM, K_d = 10 mM.

349

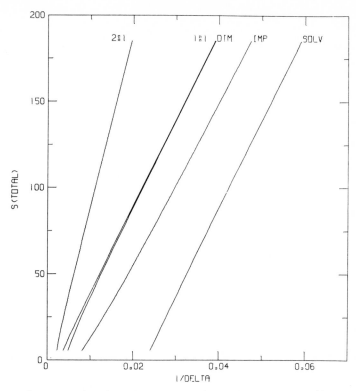

<u>Figure 2</u>   Calculated curves of S_t <u>versus</u> $1/\Delta$.
R_t = 5 mM; other parameters as in Figure 1.

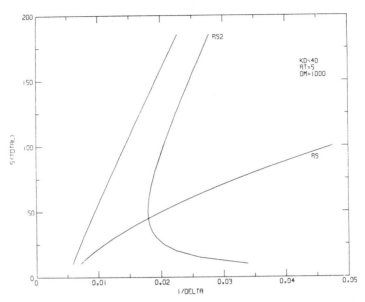

Figure 3   Calculated curves of S_t versus $1/\Delta$ giving the contribution of the RS and RS_2 species separately. Parameters used are $\Delta_M = 1000$ Hz, $R_t = 5$ mM, $K = 40$ mM.

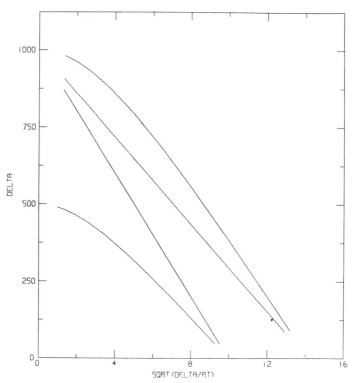

Figure 4   Calculated curves of Δ versus $\sqrt{\Delta/R_t}$, from
left to right (bottom to top): $n = 1$, $\rho = 1/2$;
$n = 1$, $\rho = 1$; $n = 2$, $\rho = 1/2$, $n = 2$, $\rho = 1$.
Parameters used are $\Delta_M = 1000$ Hz, $K = 10$ mM,
R_t in the range $0.5 - 512$ mM.

BIBLIOGRAPHY OF PAPERS RELATED TO
NMR SHIFT REAGENT CHEMISTRY

Daniel S. Dyer[*], Robert E. Sievers and Larry R. Froebe[*]
Aerospace Research Laboratories, ARL/LJ
Wright-Patterson AFB, Ohio 45433

[*]Technology Incorporated
Visiting Research Chemists

FUNDAMENTAL ASPECTS OF SHIFT
REAGENT CHEMISTRY

N. Ahmad, N. S. Bhacca, J. Selbin and J. D. Wander, J. Amer. Chem. Soc., 93, 2564 (1971). Nuclear Magnetic Resonance Spectra of Tris (2, 2, 6, 6-tetramethyl-3, 5-heptanedionato) Complexes of the Lanthanides. Temperature Dependence of Shift Reagents.

H. L. Ammon, P. H. Mazzocchi, W. J. Kopecky, Jr., H. J. Tamburin and P. H. Watts, Jr., J. Amer. Chem. Soc., 95, 1968 (1973). Lanthanide Shift Reagents. II. (a) Photochemical Ring Expansion of a β-Lactam and Product Identification Using LSR NMR Shifts and X-Ray Crystallography. (b) Probable Structure of an LSR - Substrate Complex in Solution. (c) Conformational Analysis Using LSR NMR Data.

R. von Ammon and R. D. Fischer, Angew. Chem., 11, 675 (1972). Shift Reagents in NMR Spectroscopy.

N. H. Andersen, B. J. Bottino and S. E. Smith, Chem. Commun., 1193 (1972). Evidence for Octa-co-ordination in Alcohol-Lanthanide Shift Reagent Complexes and its Implications in the Search for the Geometric Factor

Governing Observed Shifts.

J. W. ApSimon, H. Beierbeck and A. Fruchier, Can. J. Chem., 50, 2725 (1972). Conversion of Nonlinear Shift Reagent Plots to Linear Plots.

J. W. ApSimon, H. Beierbeck and A. Fruchier, Can. J. Chem., 50, 2905 (1972). Automatic Sorting of Signals in Shift Reagent Spectra.

M. K. Archer, D. S. Fell and R. W. Jotham, Inorg. Nucl. Chem. Lett., 7, 1135 (1971). Self-Induced Pseudo-Contact Shifts in Tris(2, 2, 6, 6-Tetramethyl-3, 5 Heptadionato-) Praeseodymium Dimer, $Pr_2(thd)_6$, and its Europium Homologue.

I. Armitage, G. Dunsmore, L. D. Hall and A. G. Marshall, Chem. Commun., 1281 (1971). Calculation of Binding Constants and Bound Chemical Shifts for the Association of Lanthanide Shift Reagents with Organic Substrates.

I. Armitage, G. Dunsmore, L. D. Hall and A. G. Marshall, Can. J. Chem., 50, 2119 (1972). Evaluation of the Binding Constants, Bound Chemical Shifts, and Stoichiometry of Lanthanide -Substrate Complexes.

I. Armitage, G. Dunsmore, L. D. Hall and A. G. Marshall, Chem. Ind. (London), 79 (1972). Determination of Stoichiometry for Binding of Organic Substrates to Lanthanide Shift Reagents.

C. Beaute, S. Cornuel, D. Lelandais, N. Thoai and Z. Wolkowski, Tetrahedron Lett. 1099 (1972). Temperature Dependence of NMR Shifts Induced by Tris(dipivalomethanato)ytterbium.

R. D. Bennett and R. E. Schuster, Tetrahedron Lett. 673 (1972). Lanthanide NMR Shift Reagents: An Unexpected

Temperature Effect.

N. S. Bhacca, J. Selbin and J. D. Wander, J. Amer. Chem. Soc., 94, 8719 (1972). NMR Spectra of 1:1 Adducts of 1, 10-Phenanthroline and α, α'-Bipyridyl with Tris (2, 2, 6, 6-tetramethylheptane-3, 5-dionato) Complexes of the Lanthanides.

N. S. Bhacca and J. D. Wander, Chem. Commun. 1505 (1971). Aromatic Proton Magnetic Resonances that are not Shifted by Eu(dpm)$_3$.

E. R. Birnbaum and T. Moeller, J. Amer. Chem. Soc., 91, 7274 (1969). Observations on the Rare Earths. LXXXII. NMR and Calorimetric Studies of Complexes of the Tripositive Ions with Substituted Pyridine Molecules.

B. Bleaney, J. Magnetic Res., 8, 91 (1972). NMR Shifts in Solution Due to Lanthanide Ions.

N. Bloembergen and W. C. Dickinson, Phys. Rev., 79, 179 (1950). On the Shift of the Nuclear Magnetic Resonance in Paramagnetic Solutions.

A. K. Bose and R. Brambilla, Abstract PHYS 72, 164th ACS National Meeting, New York, N. Y., Aug. 27-Sept. 1, 1972. Unusual Lanthanide-Induced Shifts in CMR Spectra.

J. Bouquant and J. Chuche, Tetrahedron Lett., 2337 (1972). Reactifs Lanthanidiques de Deplacement Chimique I - Methode Simple de Determination de K et Δ.

J. Bouquant and J. Chuche, Tetrahedron Lett. 493 (1973). NMR Lanthanide Shift Reagents. II. Solvent Effects.

C. Brecher, H. Samelson and A. Lempicki, J. Chem. Phys., 42, 1081 (1965). Lasers Phenomena in Europium

Chelates. III. Spectroscopic Effects of Chemical Composition and Molecular Structure.

J. M. Briggs, G. P. Moss, E. W. Randall and K. D. Sales, Chem. Comm., 1180 (1972). Pseudo-contact Contributions to Lanthanide-Induced NMR Shifts.

R. Caple, D. K. Harris and S. C. Kuo, J. Org. Chem., 38, 381 (1973). Nature of Lanthanide-Induced Shifts. Detection of the Angular Dependency Factor.

R. Caple and S. C. Kuo, Tetrahedron Lett. 4413 (1971). Angular Dependency of Lanthanide-Induced Shifts in Proton NMR Spectra. Analysis of Rigid Bicyclic Ethers.

A. F. Cockerill and D. M. Rackham, Tetrahedron Lett. 5149 (1970). Quantitation of the Chemical Shifts Induced by Tris(dipivalomethanato) Europium(III) in the PMR Spectra of Hydroxyadamantanes and Cyclopentanol.

D. R. Crump, J. K. M. Sanders, and D. H. Williams, Tetrahedron Lett., 4419 (1970). Evaluation of some Tris-(dipivalomethanato) Lanthanide Complexes as Paramagetic Shift Reagents.

R. J. Cushley, D. R. Anderson and S. R. Lipsky, Chem. Commun. 636 (1972). An Upfield Carbon - 13 Shift Induced by Tris(dipivalomethanato)europium.

R. E. Davis and M. R. Willcott, J. Amer. Chem. Soc., 94, 1744 (1972). Interpretation of the Pseudocontact Model for NMR Shift Reagents. II. Significance Testing on the Agreement Factor R.

R. E. Davis and R. Willcott, III, Abstracts, 13th Experimental NMR Conference, Pacific Grove, Calif., May 1972, 47. A Method for the Interpretation of the Pseudocontact Model for NMR Shift Reagents.

P. V. Demarco, B. J. Cerimele, R. W. Crane and A. L. Thakkar, Tetrahedron Lett., 34, 3539 (1972). Time-Averaged Solution Geometries for Ligand Europium Complexes: The Significance of Functional Group Rotation, Basicity, and Steric Environment.

P. V. Demarco, T. K. Elzey, R. B. Lewis and E. Wenkert, J. Amer. Chem. Soc., 92, 5734 (1970). Paramagnetic Induced Shifts in Proton Magnetic Resonance Spectra of Alcohols Using Tris(dipivalomethanato) europium(III).

D. A. Deranleau, J. Amer. Chem. Soc., 91, 4044 (1969). Theory of the Measurement of Weak Molecular Complexes. I. General Considerations.

H. Donato, Jr. and R. B. Martin, J. Amer. Chem. Soc., 94, 4129 (1972). Dipolar Shifts and Structure in Aqueous Solutions of 3:1 Lanthanide Complexes of 2, 6-Dipicolonate.

D. R. Eaton, J. Amer. Chem. Soc., 87, 3097 (1965). The NMR of Some Paramagnetic Transition-Metal Acetylacetonates.

D. R. Eaton, W. D. Phillips and D. J. Caldwell, J. Amer. Chem. Soc., 85, 397 (1963). Configurations and Magnetic Properties of the Nickel (II) Aminotroponeimineates.

L. Ernst and A. Mannschreck, Tetrahedron Lett. 3023 (1971). Eu(dpm)$_3$ -Induced Shifts in Substituted Anilines. The Importance of Basicity and Steric Effects.

S. Farid, A. Ateya and M. Maggio, Chem. Commun. 1285 (1971). Lanthanide-Induced Shifts in H Nuclear Magnetic Resonance: the Significance of the Angle Term in the Geometric Factor of the Pseudocontact Shift.

357

B. Feibush, M. F. Richardson, R. E. Sievers and C. S. Springer, Jr., J. Amer. Chem. Soc., 94, 6717 (1972). Complexes of Nucleophiles with Rare Earth Chelates.

I. Fleming, S. W. Hanson and J. K. M. Sanders, Tetrahedron Lett. 3733 (1971). The Effect of Eu(dpm)$_3$ on the NMR Spectra of Bifunctional Compounds.

L. R. Froebe, J. J. Brooks, D. S. Dyer and R. E. Sievers, Proc. XIVth I. C. C. C., Toronto, Canada, June 22-28, 1972, p. 346. Gas Chromatographic and NMR Studies of the Interaction of Nucleophiles with Lanthanide Shift Reagents.

J. S. Ghotra, F. A. Hart, G. P. Moss and M. L. Staniforth, Chem. Commun. 113 (1973). Nature of the Species Present in Solutions of Lanthanide NMR Shift Reagents.

V. G. Gibb, J. M. Armitage, L. D. Hall and A. G. Marshall, J. Amer. Chem. Soc., 94, 8919 (1972). Variation of Stoichiometry with Solvent in a Lanthanide NMR Shift Reagent Complex.

L. Goldberg and W. M. Ritchey, Spectroscopy Lett., 5, 201 (1972). A Quantitative Study of Effects of Rare Earth Chelates on NMR Chemical Shifts: I. Treatment of 1:1 Complexes and Chemical Shifts of the Pure Complexes.

J. Goodisman and R. S. Matthews, Chem. Commun. 127 (1972). The Interpretation of Lanthanide-Induced Shifts in ^{1}H Nuclear Magnetic Resonance Spectra.

A. M. Grotens and J. Smid, Abstract, ORGN 110, 163rd National ACS Meeting, Boston, 1972. Lanthanide Induced NMR Contact Shifts in Polyglycoldimethylethers.

H. Hart and G. M. Love, Tetrahedron Lett. 625 (1971). Coordination Sites and the Use of Chemical Shift Reagents in Polyfunctional Molecules.

G. E. Hawkes, D. Leibfritz, D. W. Roberts and J. D. Roberts, J. Amer. Chem. Soc., 95, 1659 (1973). NMR Shift Reagents. The Question of the Orientation of the Magnetic Axis in Lanthanide-Substrate Complexes.

G. E. Hawkes, C. Marzin, S. R. Johns and J. D. Roberts, J. Amer. Chem. Soc., 95, 1661 (1973). NMR Shift Reagents. Quantitative Estimates of Contact Contributions to Lanthanide-Induced Chemical-Shift Changes for exo-Norbornylamine.

T. Heigl and G. K. Mucklow, Tetrahedron Lett. 649 (1973). Position of Lanthanide Ion in NMR Shift Reagent-Ligand Chelates.

C. C. Hinckley, J. Amer. Chem. Soc., 91, 5160 (1969). Paramagnetic Shifts in Solutions of Cholesterol and the Dipyidine Adduct of Tris(dipivalomethanato)europium(III). A Shift Reagent.

C. C. Hinckley, M. R. Klotz and F. Patil, J. Amer. Chem. Soc., 93, 2417 (1971). Applications of Rare Earth NMR Shift Reagents. III. Graphical Analysis of Paramagnetic Shifts for Systems Having Two Coordination Sites. Testosterone and 17-α-Methyltestosterone.

M. Holik, Chemicke Listy, 66, 449 (1972), Lanthanide Shift Reagents in NMR Spectroscopy.

C. L. Honeybourne, Tetrahedron Lett. 1095 (1972). Concerning the Location of the Principal Magnetic Axis in Lanthanide Shift-Reagent Adducts.

W. DeW. Horrocks, R. C. Taylor and G. N. Lamar, J. Amer. Chem. Soc., 86, 3031 (1964). Isotropic PMR Shifts in π-Bonding Ligands Coordinated to Paramagnetic Ni(II) and Co(II) Acetylacetonates.

359

W. DeW. Horrocks, Inorg. Chem., 9, 690 (1970). Evaluation of Dipolar NMR Shifts.

W. DeW. Horrocks, and J. P. Sipe, III, J. Amer. Chem. Soc., 93, 6800 (1971). Lanthanide Shift Reagents. A Survey.

W. DeW. Horrocks, and J. P. Sipe, III, Abstract, INORG 76, 162nd ACS National Meeting, Washington, D. C., September 13-17, 1971. Lanthanide Shift Reagents. Mechanism and Structure.

W. DeW. Horrocks and J. P. Sipe, III, Abstracts, 13th Experimental NMR Conference, Pacific Grove, Calif., May 1972, 32. Paramagnetic Anisotropy and Dipolar Shifts in Shift Reagent Systems.

W. DeW. Horrocks, and J. P. Sipe, III, Abstract, INORG 94, 163rd National A. C. S. Meeting, Boston, April 1972. Lanthanide Shift Reagents. Evaluation of NMR Dipolar Shifts from Single Crystal Magnetic Studies of Some Adducts.

W. DeW. Horrocks, and J. P. Sipe, III, Science, 177, 994 (1972). Lanthanide Complexes as Nuclear Magnetic Resonance Structural Probes: Paramagnetic Anisotropy of Shift Reagent Adducts.

H. Huber, Tetrahedron Lett., 34, 3559 (1972). Über Die Lage Der Magnetischen Hauptachse in Zylindersymmetrischen Komplexen von Lanthaniden-Verschiebungsreagenzien.

H. Huber and C. Pascual, Helv. Chim. Acta, 54, 913 (1971). Über den Einfluss des Kontaktterms auf die Lanthaniden-Verschiebung der NMR-Signale Aromatischer Stickstoffheterocyclen.

H. Huber and J. Seelig, Helv. Chim. Acta, $\underline{55}$, 135 (1972). On the Complex Formation of Tris(dipivalomethanato)europium with Pyridine.

S. Jaakkola and R.W. Jotham, Inorg. Nucl. Chem. Lett., $\underline{8}$, 639 (1972). The Magnetic Susceptibility to 4.2 K of Tris(2, 2, 6, 6-Tetramethyl-3, 5-Heptadionato-) Praseodymium Dimer, $Pr_2(thd)_6$, and its Europium Homologue.

B. F. G. Johnson, J. Lewis, P. McArdle and J. R. Norton, Chem Commun. 535 (1972). Contact Effects in NMR Spectra with Lanthanide Shift Reagents.

S. R. Johns, R. A. Smith, G. E. Hawkes and J. D. Roberts, Proc. Nat. Acad. Sci. USA, $\underline{70}$, 939 (1973). NMR Shift Reagents: Evidence for Different Mechanisms of Contact Contributions to ^{13}C Shifts Induced by Nickel and Cobalt Acetylacetonates with Saturated Amines and Alcohols.

D. R. Kelsey, J. Amer. Chem. Soc., $\underline{94}$, 1764 (1972). NMR Paramagnetic Shift Reagents. The Use of Internal Protons as Standards for Structural Determinations. A Method for Determination of Complexation Equilibrium Constants.

F. Lefevre and M. L. Martin, Org. Mag. Res., $\underline{4}$, 737 (1972). Utilization des Complexes de Terres Rares en RMN.

W. B. Lewis, J. A. Jackson, J. F. Lemons, and H. Taube, J. Chem. Phys., $\underline{36}$, 694 (1962). Oxygen-17 NMR Shifts in Aqueous Solutions of Rare-Earth Ions.

C. Marzin, D. Leibfritz, G. E. Hawkes and J. D. Roberts, Proc. Nat. Acad. Sci. USA, $\underline{70}$, 562 (1973). NMR Shift Reagents: Abnormal ^{13}C Shifts Produced by Complexation of Lanthanide Chelates with Saturated Amines and n-Butyl Isocyanide.

J. E. Maskasky and M. E. Kenney, Abstract, INORG 77, 162nd ACS National Meeting, Washington, D. C., September 13-17, 1971. NMR Shift Reagents Based on Porphyrin Rings.

J. E. Maskasky and M. E. Kenney, J. Amer. Chem. Soc., 93, 2060 (1971). An Alternative to Donor-Acceptor Pseudocontact-Shift NMR Shift Reagents.

J. E. Maskasky and M. E. Kenney, J. Amer. Chem. Soc., 95, 1443 (1973). Macrocyclic NMR Shift Reagents.

P. H. Mazzocchi, H. L. Ammon and C. W. Jameson, Tetrahedron Lett. 573 (1973). Lanthanide Shift Reagents III. Errors Resulting from the Neglect of the Angle Dependence.

H. M. McConnell and R. E. Robertson, J. Chem. Phys., 29, 1361 (1958). Isotropic Nuclear Resonance Shifts.

H. M. McConnell and D. B. Chesnut, J. Chem. Phys., 28, 107 (1958). Theory of Isotropic Hyperfine Interactions in π-Electron Radicals.

G. A. Neville, Org. Mag. Res., 4, 633 (1972). Characterization of Di-and Trimethoxybenzaldehydes by Analysis of Deshielding Gradients Obtained with Eu(fod)$_3$ Shift Reagent.

M. R. Peterson, Jr. and G. H. Wahl, Jr., J. Chem. Ed., 49, 790 (1972). Lanthanide NMR Shift Reagents: A Powerful New Stereochemical Tool.

W. D. Phillips, C. E. Looney and C. K. Ikeda, J. Chem. Phys., 27, 1435 (1957). Influence of Some Paramagnetic Ions on the Magnetic Resonance Absorption of Alcohols.

A. J. Rafalski, J. Barciszewski and M. Wiewiorowski, Tetrahedron Lett. 2829 (1971). Calculation of Lanthanide

NMR Shift Dependence on Proton Distance. An Analytical Approach.

J. Reuben and D. Fiat, Chem. Commun. 729 (1967). Proton Chemical Shifts in Aqueous Solutions of the Rare-Earth Ions as an Indicator of Complex Formation.

J. Reuben and D. Fiat, J. Chem. Phys., 47, 5440 (1967). Non-Curie Behavior of Isotropic Nuclear Resonance Shifts.

J. Reuben and D. Fiat, J. Chem. Phys., 51, 4909 (1969). NMR Studies of Solutions of the Rare-Earth Ions and Their Complexes. III. Oxygen-17 and Proton Shifts in Aqueous Solutions and the Nature of Aquo and Mixed Complexes.

J. Reuben and D. Fiat, J. Chem. Phys., 51, 4918 (1969). NMR Studies of Solutions of the Rare-Earth Ions and Their Complexes. IV. Concentration and Temperature Dependence of the Oxygen-17 Transverse Relaxation in Aqueous Solutions.

J. Reuben and J.S. Leigh, Jr., J. Amer. Chem. Soc., 94, 2789 (1972). Effects of Paramagnetic Lanthanide Shift Reagents on the Proton Magnetic Resonance Spectra of Quinoline and Pyridine.

R.E. Rondeau and R.E. Sievers, J. Amer. Chem. Soc., 93, 1522 (1971): Abstracts, 12th Experimental Nuclear Magnetic Resonance Conference, Gainesville, Fla., Feb 18-20, 1971; New Superior Paramagnetic Shift Reagents for NMR Spectral Clarification.

K. Roth, M. Grosse and D. Rewicki, Tetrahedron Lett. 435 (1972). Stochiometrie und Bestandigkeit von Komplexen mit NMR Shift Reagenzien.

J. K. M. Sanders, S. W. Hanson and D. H. Williams, J. Amer. Chem. Soc., 94, 5325 (1972). Paramagnetic Shift Reagents. The Nature of the Interactions.

J. K. M. Sanders and D. H. Williams, J. Amer. Chem. Soc., 93, 641 (1971). Tris(dipivalomethanato)europium. A Paramagnetic Shift Reagent for Use in NMR Spectroscopy.

J. K. M. Sanders and D. H. Williams, Tetrahedron Lett. 2813 (1971). Evidence for Contact and Pseudo-Contact Contributions in Lanthanide ^{31}P NMR Shifts.

J. K. M. Sanders and D. H. Williams, Nature, 240, 385 (1972). Shift Reagents in NMR Spectroscopy.

S. Sato, Jap. J. App. Phys., 10, 902 (1971). Some Properties of Europium β-diketone Chelates. II (Polarized Spectra and Ligand Field Symmetrics).

G. P. Schiemenz, Tetrahedron, 29, 741 (1973). Ionen-VIII. das Tetra-phenyloborat-ion als Kernresonanzverschiebungsreagens bei Quärtaren Ammonium-Kationen.

D. Schwendiman and J. I. Zink, Inorg. Chem., 11, 3051 (1972). Solvent Effects on the Isotropic Shifts and Magnetic Susceptibility of the Shift Reagent Tris(dipivaloyl-methanato)europium(III).

M. B. Shambhu and G. A. Digenis, 24th Southeastern Regional A. C. S. Meeting, Birmingham, Alabama, Nov. 2, 1972. NMR Studies on Complexation of Eu(dpm)$_3$ at Two Sites, Amide Carbonyl and Hetero Nitrogen.

B. L. Shapiro, J. R. Hlubucek, G. R. Sullivan and L. F. Johnson, J. Amer. Chem. Soc., 93, 3281 (1971). Lanthanide-Induced Shifts in Proton NMR Spectra. I. Europium-Induced Shifts to Higher Fields.

B. L. Shapiro, M. D. Johnston, Jr., A. D. Godwin, T. W. Proulx and M. J. Shapiro, Tetrahedron Lett. 3233 (1972). Concerning the Relative Shifting Abilities of Eu(dpm)$_3$ and Eu(fod)$_3$.

B. L. Shapiro, M. D. Johnston, Jr., R. L. R. Tosns, J. Amer. Chem. Soc., 94, 4381 (1972). Lanthanide-Induced Changes in Proton Spin-Spin Coupling Constants.

B. L. Shapiro and M. D. Johnston, Jr., Abstracts, 13th Experimental NMR Conference, Pacific Grove, Calif., May 1972, 38. Lanthanide Shift Reagents: Some Complications and a New Effect.

B. L. Shapiro and M. D. Johnston, Jr., J. Amer. Chem. Soc., 94, 8185 (1972). Lanthanide-Induced Shifts in Proton NMR Spectra. III. Lanthanide Shift Reagent-Substrate Equilibria.

B. L. Shapiro, M. J. Shapiro, A. D. Godwin and M. D. Johnston, Jr., J. Mag. Res., 8, 402 (1972). Lanthanide-Induced Effects in Proton NMR Spectra. VIII. "Scavenging" Effects-A Problem and a Solution.

S. P. Sinha and R. D. Green, Spectroscopy Lett., 4, 399 (1971). NMR Studies of Lanthanide(III) Complexes. High Field Shifts in Complexes of 1,10-Phenanthroline.

J. Skolik, J. Barciszewski, A. J. Rafalski and M. Wiewiorowski, Bull. Acad. Pol. Sci., Ser. Sci. Chim., 19, 599 (1971). Europium Shifts in the Lactam Series. II. Infrared Spectroscopic Study of Shift Reagent Influence on Substrate Conformation.

L. Tomić, Z. Majerski, M. Tomić, and D. E. Sunko, Chem. Commun., 719 (1971). Temperature and Concentration Dependence of the Paramagnetic Induced Shifts in Proton Magnetic Spectroscopy.

G.H. Wahl, Jr. and M.R. Peterson, Jr., Chem. Commun. 1167 (1970). On the Mechanism of Deshielding of the Tris(dipivalomethanato)europium NMR Shift Reagent; Corrigenda, ibid., 1970, 1584.

S.I. Weissman, J. Amer. Chem. Soc., 93, 4928 (1971). On the Action of Europium Shift Reagents.

M.R. Willcott, III, R.E. Lenkinski and R.E. Davis, J. Amer. Chem. Soc., 94, 1742 (1972). Interpretation of the Pseudocontact Mode for Nuclear Magnetic Resonance Shift Reagents. I. The Agreement Factor, R.

D.E. Williams, Tetrahedron Lett. 1345 (1972). NMR Shift Reagents: Intrinsic LIS Parameters in Mixtures.

R.M. Wing, T.A. Early and J.J. Uebel, Tetrahedron Lett. 4153 (1972). A Simple Graphical Method for Analysis of Lanthanide-Induced Shifts.

T. Yonezawa, I. Morishima and Y. Ohmori, J. Amer. Chem. Soc., 92, 1267 (1970). Studies on the ^{1}H Paramagnetic Shifts for σ-Systems in Piperidine and Quinuclidine.

CHEMICAL AND PHYSICAL PROPERTIES OF SHIFT REAGENTS

M.L. Bhaumik, J. Inorg. Nucl. Chem., 27, 243 (1965). Exchange of Rare-Earth Ions in Chelate Solutions.

J.C.A. Boeyens, J. Chem. Phys., 54, 75 (1971). Ionic Interactions in Volatile Lanthanide Chelates.

K.J. Eisentraut and R.E. Sievers, J. Amer. Chem. Soc., 87, 5254 (1965). Volatile Rare Earth Chelates.

K.J. Eisentraut and R.E. Sievers, Inorg. Syn., XI, 94 (1968). Volatile Rare Earth Chelates of 2, 2, 6, 6-

tetramethylheptane-3, 5-dione.

K. J. Eisentraut and R. E. Sievers, U. S. Patents
3, 429, 904 and 3, 453, 319 (1969). Separation of Chelates
of Rare Earth Compounds and Promethium by Fractional
Sublimation (3, 429, 904). Gas Phase Chromatographic
Separation of Lanthanide Rare Earth Chelates (3,453,319).

S. J. Lyle and A. D. Witts, Inorg. Chim. Acta, 5, 481
(1971). A Critical Examination of Some Methods for the
Preparation of Tris and Tetrakis Diketonates of
Europium(III).

J. K. Przystal, W. G. Bos and I. B. Liss, J. Inorg. Nucl.
Chem., 33, 679 (1971). The Preparation and Characterization of Some Anhydrous Rare Earth Tris(acetylacetonates).

J. E. Schwarberg, D. R. Gere, R. E. Sievers and K. J.
Eisentraut, Inorg. Chem., 6, 1933 (1967). Clarification of Discrepancies in the Characterization of
Lanthanum Series Complexes of 2, 2, 6, 6-Tetramethyl-3,
5-heptanedione.

J. Selbin, N. Ahmad and N. Bhacca, Inorg. Chem., 10,
1383 (1971). Preparation and Properties of Lanthanide
Chelate Complexes.

V. Shurig, Tetrahedron Lett. 3297 (1972). Chiral Shift
Reagents for NMR Spectroscopy. A Simple and Improved
Access to Lanthanide-Tris-Chelates of d-3-TFA-
Camphor.

J. E. Sicre, J. T. Dubois, K. J. Eisentraut and R. E.
Sievers, J. Amer. Chem. Soc., 91, 3476 (1969). Volatile
Lanthanide Chelates. II. Vapor Pressures, Heats of
Vaporization, and Heats of Sublimation.

R. E. Sievers, "Coordination Chemistry", Ed. by S. Kirschner, Plenum Press, New York 1969, **pp.** 270-288. Gas Chromatographic and Related Studies of Metal Complexes.

R. E. Sievers, K. J. Eisentraut, C. S. Springer and D. W. Meek, Adv. Chem., $\underline{71}$, 141 (1968). Volatile Rare Earth Chelates of β-Diketones.

C. S. Springer, Jr., D. W. Meek and R. E. Sievers, Inorg. Chem., $\underline{6}$, 1105 (1967). Rare Earth Chelates of 1, 1, 1, 2, 2, 3, 3-Heptafluoro-7, 7-dimethyl-4, 6-octanedione.

X-RAY CRYSTALLOGRAPHIC STRUCTURAL DETERMINATIONS

J. C. A. Boeyens and J. P. R. de Villiers, J. Crystallogr. Mol. Struct., $\underline{1}$, 297 (1971). Crystal Structure of Tris (1, 1, 1, 2, 2, 3, 3-heptafluoro-7, 7-dimethyl-4, 6-octanedionato)aquolutetium(III), $Lu(fod)_3 \cdot H_2O$.

R. E. Cramer and K. Seff, Chem. Commun. 400 (1972). Crystal and Molecular Structure of a NMR Shift Reagent, the Dipyridine Adduct of Tris(2, 2, 6, 6-tetramethylheptane-3, 5 - dionato)europium(III), $Eu(dpm)_3(py)_2$.

C. S. Erasmus and J. C. A. Boeyens, Acta Cryst., $\underline{B26}$, 1843 (1970). Crystal Structure of the Praseodymium β-Diketonate of 2, 2, 6, 6-Tetramethyl-3, 5-heptanedione, $Pr_2(thd)_6$.

C. S. Erasmus and J. C. A. Boeyens, J. Crystallogr. Mol. Struct., $\underline{1}$, 83 (1971). Crystal Structure of Tris (2, 2, 6, 6-tetramethyl-3, 5-heptanedionato) aquodysprosium(III), $Dy(thd)_3 \cdot H_2O$.

W. DeW. Horrocks, J.P. Sipe, III and J.R. Luber, J. Amer. Chem. Soc., 93, 5258 (1971). Lanthanide Shift Reagents. The X-ray Structure of the Eight-Coordinate Bis-4-Picoline Adduct of 2, 2, 6, 6-Tetramethylheptane-3, 5-dionatoholmium, $Ho(dpm)_3(4-pic)_2$.

M. F. Richardson, P.W.R. Corfield, D. E. Sands, R. E. Sievers, Inorg. Chem., 9, 1632 (1970). The Crystal and Molecular Structure of the Acetylacetonimine Adduct of Ytterbium Acetylacetonate, $Yb(C_5H_7O_2)_3$ $(CH_3COCH = C(NH_2)CH_3)$.

J.J. Uebel and R.M. Wing, J. Amer. Chem. Soc., 94, 8910 (1972). Lanthanide Shift Reagents. X-Ray Structure of the Seven-Coordinate 3, 3-Dimethylthietane 1-Oxide (1) Complex with Tris(dipivalomethanato)europium(III), $Eu(dpm)_3 \cdot (1)$, and Its Implication on Pseudocontact Shift Calculations.

J.P.R. de Villiers and J.C.A. Boeyens, Acta Cryst., B27, 692 (1971). The Crystal and Molecular Structure of the Hydrated Praseodymium Chelate of 1, 1, 1, 2, 2, 3, 3-Heptafluoro-7, 7-dimethyl-4, 6-octanedione, $Pr_2(fod)_6 \cdot 2H_2O$.

J.P.R. de Villiers and J.C.A. Boeyens, Acta Cryst., B 28, 2335 (1971). Crystal Structure of tris(2, 2, 6, 6-tetramethyl-heptane-3, 5-dionato)erbium(III).

S.J.S. Wasson, M.S. Thesis, U. of Ky., The Crystal and Molecular Structure of 3-Methylpyridine-tris (2, 2, 6, 6-tetramethyl-3, 5-heptanedionato)Lu(III), 1972

S.J.S. Wasson, D. E. Sands and W. F. Wagner, Inorg. Chem., 12, 187 (1973). Crystal and Molecular Structure of 3-Methylpyridine-Tris(2, 2, 6, 6-tetramethyl-3, 5-heptanedionato)lutetium(III).

369

GENERAL APPLICATIONS OF SHIFT REAGENTS

O. Achmatowicz, Jr., A. Ejchart, J. Jurczak, L. Kozerski and J. St. Pyrek, Chem. Commun. 98 (1971). Confirmation of the Structure of a New Diterpene Trachyloban-19-ol, by Tris(dipivaloylmethanato)europium shifted NMR Spectroscopy.

K.K. Andersen and J.J. Uebel, Tetrahedron Lett. 5253 (1970). Use of Tris(dipivalomethanato)europium(III) as a Nuclear Resonance Chemical Shift Reagent in the Analysis of the Spectra of Some Sulfoxides.

J.W. ApSimon and H. Beierbeck, Chem. Commun. 172 (1972). A Method for Compensating for Experimental Errors in Shift Reagent Work.

J.W. ApSimon and H. Beierbeck, Tetrahedron Lett. 581 (1973). Lanthanide Shift Reagents. A Novel Method for Fitting the Pseudocontact Shielding Equation to Experimental Induced Shifts.

J.W. ApSimon, H. Beierbeck and A. Fruchier, J. Amer. Chem. Soc., 95, 939 (1973). On the Stoichiometry of the Lanthanide Shift Reagent-Substrate Complex.

W.L.F. Armarego, T.J. Batterham and J.R. Kershaw, Org. Mag. Res., 3, 575 (1971). Tris(dipivaloylmethanato)europium-Induced Shifts of Proton Resonances in π-Deficient Nitrogen Heterocycles.

I. Armitage, J.R. Campbell and L.D. Hall, Can. J. Chem., 50, 2139 (1972). Applications of Lanthanide Shift Reagents to the Identification of ^{13}C Resonances.

I. Armitage and L.D. Hall, Chem. Ind. (London), 1537 (1970). Some Observations on the Chemical Shift Changes Induced by Europium Tris(dipivaloylmethane).

I. Armitage and L. D. Hall, Can. J. Chem., $\underline{49}$, 2770 (1971). Novel Chemical Shift Changes of Carbohydrates Induced by Lanthanide Shift Reagents: Some Experimental Optimizations.

R. L. Atkins, D. W. Moore and R. A. Henry, J. Org. Chem., $\underline{38}$, 400 (1973). ^{1}H NMR Structure Elucidation of Substituted Isoquinolines by Means of $Eu(fod)_3$-Induced Paramagnetic Shifts.

J. Bargon, J. Amer. Chem. Soc., $\underline{95}$, 941, (1973). Chemically Induced Dynamic Nuclear Polarization in the Presence of Paramagnetic Shift Reagents.

C. Beaute, Z. W. Wolkowski and N. Thoai, Chem. Commun. 700 (1971). Tris(dipivaloylmethanato) ytterbium-Induced Shifts in the ^{1}H NMR Spectra of Some Nitrogen Compounds.

C. Beaute, Z. W. Wolkowski and N. Thoai, Tetrahedron Lett. 817 (1971). Tris(dipivalomethanato)ytterbium-Induced Shifts in PMR--Amines.

J. P. Bégué, Bull. Soc. Chim France, 2073 (1972). Deplacements Paramagnetics Induits par les Chélates de Terres Rares en RMN: Utilisation in Chimie Organique.

P. Bélanger, C. Freppel. D. Tizané and J. C. Richer, Chem. Commun. 266 (1971). Lanthanide-Induced Shifts in NMR Spectroscopy: Application to Ketones.

B. Birdsall, J. Feeney, J. A. Glasel, R. J. P. Williams and A. V. Xavier, Chem. Commun. 1473 (1971). A Method of Assigning ^{13}C NMR Spectra Using Europium (III) Ion-Induced Pseudocontact Shifts and C-H Heteronuclear Spin Decoupling Techniques.

B. Bleaney, C. M. Dobson, B. A. Levine, R. J. P.
Williams and A. V. Xavier, Chem. Commun. 791 (1972).
Origin of Lanthanide NMR Shifts and Their Uses.

R. A. Bletton, G. F. H. Green and J. E. Page, Chem.
Commun. 1134 (1972). Effect of Lanthanide Shift Reagents on the NMR Spectra of Secondary Amides.

F. Bohlmann and C. Zdero, Tetrahedron Lett. 621
(1972). Zwei Neue Sesquiterpen-Lactone aus Lidbeckia
Pectinata Berg. und Pentzia Elegans DC.

F. Bohlman and C. Zdero, Tetrahedron Lett. 851
(1972). Ein Neues Furansesquiterpen aus Phyma-
spermum Parvifolium.

A. F. Bramwell, G. Riezebos and R. D. Wells,
Tetrahedron Lett. 2489 (1971). Correlations of NMR
Spectral Parameters with Structure: Benzylic Coupling
in Substituted Pyrazines.

A. F. Bramwell and R. D. Wells, Tetrahedron, $\underline{28}$, 4155
(1972). The NMR Spectra of Pyrazines. Benzylic
Coupling in Substituted Methylpyrazines.

J. Briggs, G. H. Frost, F. A. Hart, G. P. Moss and
M. L. Staniforth, Chem. Commun. 749 (1970).
Lanthanide-Induced Shifts in NMR Spectroscopy. Shifts
to High Field.

J. Briggs, F. A. Hart and G. P. Moss, Chem. Commun.
1506 (1970). The Application of Lanthanide-Induced Shifts
to the Complete Analysis of the Borneol NMR Spectrum.

J. Briggs, F. A. Hart, G. P. Moss and E. W. Randall,
Chem. Commun. 364 (1971). A Ready Method of Assignment for ^{13}C NMR Spectra: The Complete Assignment
of the ^{13}C Spectrum of Borneol.

A. van Bruijnsvoort, C. Kruk, E. R. de Waard and H. O. Huisman, Tetrahedron Lett. 1737 (1972). Elucidation of Relative Configuration of Thiadecalones with $Eu(dpm)_3$; Preferred Site of Complexation.

D. G. Buckley, G. H. Green, E. Ritchie and W. C. Taylor, Chem. Ind. (London), 298 (1971). Application of the Tris-(dipivalomethanato)europium NMR Shift Reagent to the Study of Triterpenoids.

J. R. Bull and A. J. Hodgkinson, Tetrahedron, 28, 3969 (1972). Steroidal Analogues of Unnatural Configuration - VI. A-Ring Transformations of 4,4,14 α -Trimethyl-19 (10→9 β)ABEO-10 α -PREGN-5-ENE-3,11, 20-Trione.

C. A. Burgett and P. Warner, J. Magnetic Res., 8, 87 (1972). A New Lanthanide Paramagnetic Shift Reagent Containing Fully Fluorinated Side Chains.

H. Burzynska, J. Dabrowski and A. Krowczynski, Bull. Acad. Pol. Sci., Ser. Sci. Chim., 19, 587 (1971). Europium - and Praseodymium-Induced Shifts in the PMR Spectra of Amines.

R. F. Butterworth, A. G. Pernet and S. Hanessian, Can. J. Chem., 49, 981 (1971). Utility of Shift Reagents in NMR Studies of Polyfunctional Compounds: N-and O-Acetylated Carbohydrates and Nucleosides.

P. -F. Casals and G. Boccaccio, Tetrahedron Lett. 1647 (1972). Photodimers de la Dimethyl-4,4 Cyclohexene-2 one. Syntheses de Bis α -Enones de la Serie du Tricyclo $(6, 4, 0, 0^{2,7})$ Dodecane.

C. P. Casey and C. R. Cyr, J. Amer. Chem. Soc., 93, 1280 (1971). Hydroformylation of 3-Methyl-1-hexane-3-d_1. Evidence Against Direct Formylation of a Methyl Group in the "Oxo" Reaction.

P. Caubere and J.J. Brunet, Tetrahedron, <u>28</u>, 4859
(1972). Condensation en Milieu Aprotique des Enolates
de Cetones sur le Chloro-1-cyclohexene en Presence de
Bases - III. Cas de Cetones Aromatiques.

A.A. Chalmers and K.G.R. Rachler, Tetrahedron Lett.
4033 (1972). ^{13}C and ^{1}H NMR Spectra of Quinoline in the
Presence of Lanthanide Shift Reagents.

M. Christl, H.J. Reich, and J.D. Roberts, J. Amer.
Chem. Soc., <u>93</u>, 3463 (1971). NMR Spectroscopy.
Carbon-13 Chemical Shifts of Methylcyclopentanes,
Cyclopentanols, and Cyclopentyl Acetates.

R.M. Claramunt, J. Elguero and R. Jacquier, Org. Mag.
Res., <u>3</u>, 595 (1971). Etudes RMN en Serie Hetero-
cyclique.

G.L. Closs, Special Lectures presented at the XIIIrd
IUPAC Congress, Volume <u>4</u>, 19 (1971), Butterworth,
(London).

A.F. Cockerill, N.J.A. Gutteridge, D.M. Rackham and
C.W. Smith, Tetrahedron Lett., <u>30</u>, 3059 (1972). For-
mation of Bis(1-Amino-2-Naphthalene) Trisulphide by
Fusion of 1-Naphthylamine and Sulphur.

A.F. Cockerill and D.M. Rackham, Tetrahedron Lett.
5153 (1970). Elucidation of the PMR Spectrum of 2-
Hydroxy-1-(2-Hydroxyethyl)-Adamantane in Presence of
Tris(dipivalomethanato)europium(III).

R.M. Cory and A. Hassner, Tetrahedron Lett. 1245
(1972). Proton Magnetic Resonance Assignments in
Dichloroketene-Olefin Adducts by Lanthanide-Induced
Shifts.

L. Crombie, D. A. R. Findley and
D. A. Whiting, Tetrahedron Lett. 4027 (1972). Lantha-
nide-Induced Chemical Shifts in Natural Cyclopropanes;
Stereochemistry of Chrysanthemyl and Presqualene
Alcohols and Esters.

N. S. Crossley, J. Chem. Soc. C., 2491 (1971). Seco-
steroids. Part 1. 15, 16-Secoprogesterone.

D. R. Crump, J. K. M. Sanders and D. H. Williams,
Tetrahedron Lett. 4949 (1970). Some Applications of
Paramagnetic Shift Reagents in Organic Chemistry.

B. D. Cuddy, K. Treon and B. J. Walker, Tetrahedron
Lett. 4433 (1971). The Use of Lanthanide Shift Reagents in Structural Studies of Phosphine Oxides.

D. W. Darnall, E. R. Birnbaum, J. E. Gomez and G. E.
Smolka, Proceedings of the 9th Rare Earth Research
Conference, Ed. by P. E. Field, Blacksburg, Va., Oct.
10-14, 1971, p. 278. Rare Earth Metal Ions as Probes
of Calcium Ion Binding Sites in Proteins.

P. V. Demarco, T. K. Elzey, R. B. Lewis and E.
Wenkert, J. Amer. Chem. Soc., 92, 5737 (1970).
Tris(dipivalomethanato)europium(III). A Shift Reagent
for Use in the Proton Magnetic Resonance Analysis of
Steroids and Terpenoids.

J. C. Duggan, W. H. Urry and J. Schaefer, Tetrahedron
Lett. 4197 (1971). Carbon-13 Pseudo-Contact Shifts in
Structure Determination: The Hemiketal Dimer from 2-
hydroxy-2-methylcyclobutanone.

L. Ernst, Nachr. Chem. Techni., 18, 459 (1970).
Paramagnetische Lanthaniden-Komplexe als "Verchie-
bungsreagentien" in der NMR Spektroskopie.

J. W. Faller and G. N. LaMar, Tetrahedron Lett. 1381 (1973). Chemical Spin Decoupling and Lanthanide Reagents.

S. Farid and K. -H. Scholz, J. Org. Chem., 37, 481 (1972). Ring Expansion of Hydroxyoxetanes to Di-hydrofurans.

W. Fink, Helv. Chim. Acta, 54, 1304 (1971). Tetrakis(triphenylphosphin)platin(0), ein Katalysator zur Selektiven Hydrosilylierung.

R. A. Fletton, G. F. H. Green and J. E. Page, Chem. Ind. (London), 167 (1972). Lanthanide -Induced Dis-placements in the NMR Spectra of Heterocyclic N-oxides.

M. I. Foreman and D. G. Leppard, J. Organometal. Chem., 31, C31 (1971). NMR Spectra of Organometallic Compounds: Use of a Lanthanide Shift Reagent.

H. E. Francis and W. F. Wagner, Org. Mag. Res., 4, 189 (1972). Induced Chemical Shifts in Organic Mole-cules. Intermediate Shift Reagents.

B. Franzus, S. Wu, W. C. Baird, Jr. and M. L. Scheinbaum. J. Org. Chem., 37, 2759 (1972). An NMR Technique for Distinguishing Isomers of 3, 5-Disubsti-tuted Nortricyclenes.

R. R. Fraser and Y. Y. Wigfield, Chem. Commun. 1471 (1970). Assignment of Configuration to Sulphoxides by Use of the "Shift Reagent".

O. A. Gansow, M. R. Willcott, and R. E. Lenkinski, J. Amer. Chem. Soc., 93, 4295 (1971). Carbon Magnetic Resonance. Signal Assignment by Alternately Pulsed NMR and Lanthanide -Induced Chemical Shifts.

M. D. Gheorghin, C. Draghici, L. Stanescu and M. Avram, Tetrahedron Lett. 9 (1973). 2-Cyclobuten-1-one Derivatives from Acetylenes and t-Butylcyanoketene.

M. Gielen, N. Goffin and J. Topart, J. Organomet. Chem., 32, C38 (1971). Organometallic Compounds XXXIII. Influence of tris(dipvaloylmethanato)-europium(III) on the PMR Spectra of, and Evidence for Intramolecular Complexation in Functionally-Substituted Organotin Compounds.

E. Gillies, W. A. Szarek and M. C. Baird, Can. J. Chem., 49, 211 (1971). Application of Paramagnetic Shift Reagents in NMR Studies of Complex Alcohols and Amines.

P. Girard, H. Kagan and S. David, Bull. Soc. Chim. Fr. 4515 (1970). Spectroscopie RMN dans la Serie des Glucides en Presence de Chelates de Terres Rares.

L. L. Graham, Abstracts, 163rd National A. C. S. Meeting, Boston, April 1972. NMR Studies of Amides Using Lanthanide Shift Reagents.

R. D. Green and S. P. Sinha, Spectroscopy Lett., 4, 411 (1971). Lanthanide Shift Reagents: Acetylacetone Complex of Yb(III) Ion.

G. Grethe, H. L. Lee, M. R. Vskokovic and A. Brossi, Helv. Chim. Acta, 53, 1304 (1971). Eine neue Synthese von Petalin.

A. M. Grotens, C. W. Hilbers and E. de Boer, Tetrahedron Lett. 2067 (1972). Isotope Effects in NMR Spectra of Polyglycoldimethylethers Complexed with Rare-Earth Shift Reagents.

A. M. Grotens, J. Smid and E. de Boer, Tetrahedron Lett. 4863 (1971). Lanthanide-Induced Shifts in Polyglycoldimethylethers.

J. E. Guillett, I. R. Peat and W. F. Reynolds, Tetra-hedron Lett. 3493 (1971). The Effect of Lanthanide Shift Reagents on the NMR Spectrum of Stereo-Regular Poly (Methyl Methacrylate).

M. Hajek, L. Vodicka, Z. Ksandr and S. Landa, Tetra-hedron Lett. 4103 (1972). NMR Analysis of Adaman-tane Derivatives -I. $Eu(dpm)_3$ as Shift Reagent for 2-Alkyl-2-adamantanols.

H. Hart and G. M. Love, J. Amer. Chem. Soc., 93, 6264 (1971). A New Degenerate Cyclopropylcarbinyl Cation.

H. Hart and G. M. Love, J. Amer. Chem. Soc., 93, 6266 (1971). A Novel Intramolecular Ketene Cyclo-addition. Functionalized Tetracyclo $(3.3.0.0^{2,8}.0^{3,6})$-octanes.

F. A. Hart, G. P. Moss, and M. L. Staniforth, Tetra-hedron Lett. 3389 (1971). Lanthanide-Induced Differ-ential Shifts in NMR Spectra of Aqueous Solutions.

F. A. Hart, J. W. Newbery and D. Shaw, Chem. Commun. 45 (1967). The ^1H NMR Spectra of Diamagnetic and of Paramagnetic Complexes of Rare-Earth Salts with an Aromatic Amine.

J. E. Herz, V. M. Rodriguez and P. Joseph-Nathan, Tetrahedron Lett. 2949 (1971). Stereospecific Inter-action of Ketals with Tris(dipivalomethanato)europium (III).

C. C. Hinckley, J. Org. Chem., 35, 2834 (1970). Appli-cations of Rare Earth NMR Shift Reagents. II. The Assignment of the Methyl Proton Magnetic Resonances of d-Camphor.

F. F. -L.Ho, Polymer Lett., 9, 491 (1971). Application of the Eu(dpm)$_3$ Chemical Shift Reagent to the Determination of the Molecular Weight of Poly(Propylene Glycol) by NMR.

W. DeW. Horrocks, D. L. Johnston and R. F. Venteicher, Proc. XIVth I. C. C. C., Toronto, Canada, June 22-28, 1972, p. 133. Lanthanide Complexes as Structural NMR Probes.

L. R. Isbrandt and M. T. Rogers, Chem. Commun. 1378 (1971). Lanthanide-Induced Shifts in ^{1}H NMR Spectroscopy: Resonance Assignments in Tertiary Amides.

Y. Kashman and O. Awerbouch, Tetrahedron, 27, 5593 (1971). Complexing of Phosphoryl-Containing Compounds by Tris(dipivalomethano)europium(III).

Y. Kashman and O. Awerbouch, Tetrahedron, 29, 191 (1973). The Cycloaddition Reaction of 3, 4-Dimethyl-1-Thio-1-Phenylphosphole with Tropone.

A. R. Katritzky and A. Smith, Tetrahedron Lett. 1765 (1971). Application of Contact Shift Reagents to the NMR Spectra of Polymers.

L. H. Keith, Tetrahedron Lett. 3 (1971). Eu(dpm)$_3$ Transannularly Induced Paramagnetic Chemical Shifts in PMR Spectra of Endrin, Dieldrin, and Photodieldrin.

C. A. Kingsbury, J. Org. Chem., 37, 102 (1972). Asymmetric Synthesis of Diastereomeric Hydroxy Sulfides, Sulfoxides, and Sulfones by Condensation and Oxidation Reactions.

H. Kjosen and S. Liaaen-Jensen, Acta Chem. Scand., 26 2185 (1972). Application of the Tris(dipivalomethanato)-europium(III) NMR Shift Reagent to Carotenoids.

P. Kritiansen and T. Ledaal, Tetrahedron Lett. 2817 (1971). Tris(dipivaloylmethanato)europium(III). -Induced PMR Shifts of Cyclic Ketones.

P. Kristiansen and T. Ledaal, Tetrahedron Lett. 4457 (1971). Lanthanide-Induced PMR Shifts of Cyclic Ketones. A Comparison of $Pr(dpm)_3$ and $Eu(dpm)_3$ as Shift Regents.

G. R. Krow and K. C. Ramey, Tetrahedron Lett. 3141 (1971). The Cope Rearrangement: Substituent Effects on Equilibria of Bridged Homotropilidenes.

F. Lafuma and M. C. Quivoron, C. R. Acad. Sci., Paris, Ser. C, 272, 2020, (1971). Etude des Spectres de RMN de Quelques Acetals en Presence de Tri-(dipivalomethanate) d'Europium.

F. Lefevre and M. L. Martin, Org. Mag. Res., 4, 737 (1972). Utilisation des Complexes de Terres Rares en RMN.

L. E. Legler, S. L. Jindal and R. W. Murray, Tetrahedron Lett. 3907 (1972). NMR Spectra of Thiosulfinates Containing Heterosteric Groups. Use of a Chemical Shift Reagent.

T. J. Leitereg, Tetrahedron Lett. 2617 (1972). Synthesis of Sesquiterpenoids related to Nootakatone--Structure Determination by NMR using Tris(dipivalomethanato)europium.

A. H. Lewin, Tetrahedron Lett. 3583 (1971). Restricted Rotation in Amides IV. Resonance Assignments in Tertiary Amides and Thioamides Utilizing NMR Shift Reagents.

K. J. Liska, A. F. Fentiman, Jr., and R. L. Foltz, Tetrahedron Lett. 4657 (1970). Use of Tris(dipivalomethanato)europium as a Shift Reagent in the Identification of 3-H-Pyrano[3, 2-f]Quinolin-3-One.

K. -T. Liu, Tetrahedron Lett. 5039 (1972). PMR Study of 2-Endo-Hydroxymethyl-5-Norbornene. The Effect of Tris(dipvalomethanato)europium on Coupling Constants.

G. Lukacs, X. Lusinchi, P. Girard and H. Kagan, Bull. Soc. Chim. Fr., 3200 (1971). Attribution en RMN des protons d'unsysteme AB par complexation avec Eu(dpm)$_3$.

A. Lus, G. Vecchio and G. Carrea, Tetrahedron Lett. 1543 (1972). The Use of Eu(dpm)$_3$ with Bifunctional Molecules. Additivity of the Induced Chemical Shift Changes in the NMR Spectrum.

L. J. Luskus and K. N. Houk, Tetrahedron Lett. 1925 (1972). Pyrazolotropones from the Cycloaddition of Diazomethane to Tropone.

G. Maier, G. Fritschi and B. Hoppe, Tetrahedron Lett. 1463 (1971). Chlorsubstituierte Cyclobutadiene-1, 2.

P. E. Manni, G. A. Howie, B. Katz and J. M. Cassady, J. Org. Chem., 37, 2769 (1972). Simplification of Epoxide and Lactone PMR Spectra Using Tris(dipivalomethanato)europium Shift Reagent.

T. J. Marks, J. S. Kristoff, A. Alich and D. F. Shriver, J. Organometal. Chem., 33, C35 (1971). Rare Earth Shift Reagents as Chemical and Structural Probes for Organometallic Compounds.

A. G. Marshall, I. Armitage and L. D. Hall, Abstracts, 13th Experimental NMR Conference, Pacific Grove, Calif., May 1972, 109. New and Improved Uses for Lanthanide NMR Chemical Shift Reagents.

J.E. Maskasky, J.R. Mooney and M.E. Kenney, J. Amer. Chem. Soc., 94, 2132 (1972). Iron (II) Phthalocyanines as NMR Shift Reagents for Amines.

P.H. Mazzocchi, H.J. Tamburin and G.R. Miller, Tetrahedron Lett. 1819 (1971). Upfield and Downfield Shifts in the NMR Spectrum of a Tris(dipivalomethanato) europium(III) Complex.

A.L. Meyer and R.B. Turner, Tetrahedron, 27, 2609 (1971). An Interesting Synthesis of 3-Methoxy-2,6-Dimethylphenethyl Alcohol.

L.W. Morgan and M.C. Bourlas, Tetrahedron Lett. 2631 (1972). Eu(fod)$_3$ -Induced Shifts in a Heteronuclear Bicyclic Amine.

G.A. Neville, Can. J. Chem., 50, 1253 (1972). Use of Eu(fod)$_3$ Shift Reagent and Solvent Effects in Structural Elucidation of Novel Isomeric Pyrimidones and Model Methoxylated Pyrimidines.

M. Okigawa, N. Kawano, W. Rahman and M.M. Dhar, Tetrahedron Lett. 4125 (1972). Paramagnetic Induced Shifts in the NMR Spectra of Flavone Compounds by Eu(fod)$_3$.

J. Paasivirta, Suomen Kemistilehti, B, 44, 131 (1971). Chemical Shift Reagents in the Study of Polycyclic Alcohols I. PMR Spectra of 2-Norbornanols and Dehydronorborneols.

J. Paasivirta, Suomen Kemistilehti, B, 44, 135 (1971). Chemical Shift Reagents in the Study of Polycyclic Alcohols II. PMR Spectra of 3-Nortricyclanol and Stereoisomeric 1-Methyl-3-Nortricyclanols.

J. Paasivirta and P.J. Malkonen, Suomen Kemistilehti, B, 44, 230 (1971). Chemical Shift Reagents in the Study

382

of Polycyclic Alcohols III. PMR Spectra of Methyl-substituted 5-Norbornen-2-ols.

T. B. Patrick and P. H. Patrick, J. Amer. Chem. Soc., 94, 6230 (1972). Reliability of Coupling Constants Obtained from Tris(dipivalomethanato)europium-Shifted Proton Magnetic Resonance Spectra.

J. Paul, K. Schogl and W. Silhan, Monatshefte für Chemie, 103, 243 (1972). The Application of Tris-(dipivalomethanato)europium in the NMR Spectroscopy of Metallocenes.

B. S. Perret and I. A. Stenhouse, U. K. At. Energy Auth., Res. Group, Rep. 1971 AERE-R 6824, 5pp. Application of Tris(dipivalomethanato)europium(III) as a Shift Reagent in NMR Spectroscopy.

P. E Pfeffer and H. L. Rothbart, Tetrahedron Lett. 2533 (1972). PMR Spectra of Triglycerides: Discrimination of Isomers with the Aid of a Chemical Shift Reagent.

D. L. Rabenstein, Anal. Chem., 43, 1599 (1971). Applications of Paramagnetic Shift Reagents in Proton Magnetic Resonance Spectrometry.

J. Reuben and D. Fiat, Chem. Commun. 729 (1967). Proton Chemical Shifts in Aqueous Solutions of the Rare-earth Ions as an Indicator of Complex Formation.

C. Reyes-Zamora and C. S. Tsai, Chem. Commun. 1047 (1971). NMR Isotropic Shifts of Europium Oxaloacetates.

E. W. Robb, P. P. Choma and E. K. Onsager, Abstract, ORGN 145, 164th ACS National Meeting, New York, N. Y., Aug. 27-Sept. 1, 1972. Lanthanide-Induced Shifts in Amino Acids.

J. D. Roberts, R. A. Smith, M. Christl, D. Leibfritz, C. Marzin, G. E. Hawkes, S. R. Johns, R. A. Cooper and D. W. Roberts, Abstracts, 13th Experimental NMR Conference, Pacific Grove, Calif., May 1972, 46. Shift Reagents in Carbon-13 NMR.

R. E. Rondeau, M. A. Berwick, R. N. Meppel and M. P. Serve, J. Amer. Chem. Soc., 94, 1096 (1972). Central Linkage Influence upon Mesomorphic and Electrooptical Behavior of Diaryl Nematics. A General Proton Magnetic Resonance Method Employing a Lanthanide Shift Reagent for Analysis of Isomeric Azoxybenzenes.

R. E. Rondeau and R. E. Sievers, Anal. Chem., accepted for publication. NMR Shift Reagents -A Guide to their Selection and Use.

J. K. M. Sanders and D. H. Williams, Chem. Commun. 422 (1970). A Shift Reagent for Use in NMR Spectroscopy. A First-order Spectrum of n-Hexanol.

M. P. Serve, R. E. Rondeau, H. M. Rosenberg, Abstract ORG 111, 162nd ACS National Meeting, Sept. 12-17, 1971, Washington, D. C. The Use of the Shift Reagent Europium(fod)$_3$ in Determining the Structures of Some 2-Oxabicyclo- (4. 2. 0) Octane Derivatives.

M. P. Serve, R. E. Rondeau and H. M. Rosenberg, J. Heterocycl. Chem., 9, 721 (1972). Application of Eu(fod)$_3$ in Elucidation of Structures of Bicyclic Ethers.

R. Seux, G. Morel and A. Foucaud, Tetrahedron Lett. 1003 (1972). Decyanuration des Succinonitriles Trisubstites. Configurations Z et E des Cinnamonitriles α, β -Dialcoyles Obtenus.

B. L. Shapiro, M. D. Johnston, Jr., and T. W. Proulx, J. Amer. Chem. Soc., 95, 520 (1973). 3-(α–Naphthyl)

5, 5-dimethylcyclohexanone and Derived Alcohols. Synthesis and Stereochemical Studies by Means of Lanthanide-Induced Proton NMR Shifts.

T. H. Siddall, III, Chem. Commun. 452 (1971). An Upfield ^{1}H NMR Shift Induced by Tris(dipivalomethanato) europium.

R. E. Sievers and R. E. Rondeau, Report No. ARL 70-0285, U.S. Air Force, 1970. NMR Spectral Chromatography: New Superior Paramagnetic Shift Reagents for Spectral Clarification.

W. B. Smith and D. L. Deavenport, J. Mag. Res., 6, 256 (1972). The Effect of Eu(dpm)$_3$ on the ^{13}C NMR Spectrum of Cholesterol.

G. E. Stolzenberg, R. G. Zaylski and P. A. Olson, Anal. Chem., 43, 908 (1971). NMR Identification of o, p-Isomers in an Ethoxylated Alkyphenol Nonionic Surfactant as Tris(2, 2, 6, 6-tetramethylheptane-3, 5-dione)europium(III) Complexes.

D. Swern and J. P. Wineburg, J. Amer. Oil Chem. Soc., 48, 371 (1971). NMR Chemical Shift Reagents. Application to Structural Determination of Lipid Derivatives.

W. A. Szarek, E. Dent. T. B. Grindley and M. C. Baird, Chem. Commun. 953 (1969). The Use of Paramagnetic Transition-Metal Ions in the Interpretation of NMR Spectra of Complex Alcohols and Amines.

M. Tada, Y. Moriyama, Y. Tanahashi and T. Takahashi, Tetrahedron Lett. 4007 (1971). 10(S)-Hydroxy-6(R), 9(S)-Oxidohexadecanoic Acid, A New Acid in Wool Fat.

A. Tangerman and B. Zwanenburg, Tetrahedron Lett. 79 (1973). Assignment of Configuration to Sulfines by Means of Shift Reagent and ASIS.

R. C. Taylor and D. B. Walters, Tetrahedron Lett. 63
(1972). Europium-Induced Shifts in the ^{1}H and ^{31}P NMR
of "Mixed" Nitrogen-Phosphorus Bidentate Ligands:
Evidence for a Eu(dpm)$_3$ -P(III) Interaction.

S. B. Tjan and F. R. Visser, Tetrahedron Lett. 2833
(1971). PMR Shift to High Field Induced by Tris(dipivalomethanato)europium.

L. Tomic, Z. Majerski, M. Tomic and D. E. Sunko,
Croat. Chim. Acta, 43, 267 (1971). Tris(Dipivalomethanato) holmium-Induced NMR Shifts.

K. Tori, Y. Yoshimura and R. Muneyuki, Tetrahedron
Lett. 333 (1971). Application of Paramagnetic Induced
Shifts in PMR Spectra to the Determination of the Posi-
tions of Deuterium Substitution in Bornanes Derived
from α -Pinene Using Tris(dipivalomethanato)europium
(III).

K. Tsukida and M. Ito, Experientia, 27, 1004 (1971).
Application of a Shift Reagent in NMR Spectroscopy of
Esters. An Approach for Simple Identification and
Simultaneous Determination of Tocopherols.

E. Vedejs and M. F. Salomon, J. Amer. Chem. Soc.,
92, 6965 (1970). Borohydride Reduction of σ -Bonded
Organopalladium Complexes in the Norbornenyl-
Nortricyclenyl System. Evidence Against a Radical
Mechanism.

D. B. Walters, Anal. Chim. Acta, 60, 421 (1972). Carbon Disulfide as a Solvent for the Application of Eu(dpm)$_3$
to NMR Spectroscopy.

H. Wamhoff, C. Materne and F. Knoll, Chem. Ber., 105,
753 (1972). Die Synthese von Furo (2. 3-e) -1. 4-dia-
zepinen und deren NMR. Analyse unter Verwendung des
Tris(dipivalomethanato)europium Komplexes.

T. M. Ward, I. L. Allcox and G. H. Wahl, Jr., Tetrahedron Lett. 4421 (1971). Lanthanide-Induced Shifts Due to Coordination at Phosphoryl and Amide Sites.

E. Wenkert and B. L. Buckwalter, J. Amer. Chem. Soc., 94, 4367 (1972). Carbon-13 NMR Spectroscopy of Naturally Occurring Substances. X. Pimaradienes.

E. Wenkert, D. W. Cochran, E. W. Hagaman, R. B. Lewis and F. M. Schell, J. Amer. Chem. Soc., 93, 6271 (1971). Carbon-13 NMR Spectroscopy with the Aid of a Paramagnetic Shift Agent.

W. Wiegrebe, J. Fricke, H. Budzikiewicz and L. Pohl, Tetrahedron, 28, 2849 (1972). Synthese eines 3-Phenylisochromans.

D. E. Williams, Abstracts, 13th Experimental NMR Conference, Pacific Grove, Calif., May 1972, 43. LIS (Lanthanide-Induced NMR Shift) Reagents: A Better Way to Use Them.

J. P. Wineburg and D. Swern, J. Amer. Oil Chem. Soc., 49, 267 (1972). NMR Chemical Shift Reagents in Structural Determination of Lipid Derivatives: II. Methyl Petroselinate and Methyl Oleate.

M. Witanowski. L. Stefaniak, H. Januszewski and Z. W. Wolkowski, Chem. Commun. 1573 (1971). Selection of Dipivaloylmethanate Chelates as Shift Reagents for Nitrogen-14 NMR.

M. Witanowski, L. Stefaniak, H. Januszewski and Z. W. Wolkowski, Tetrahedron Lett. 1653 (1971). Effect of Yb(dpm)$_3$ and Eu(dpm)$_3$ on Nitrogen-14 Magnetic Resonance.

T. A. Wittstruck, J. Amer. Chem. Soc., 94, 5130 (1972). Analysis of Steroid NMR Spectra Using Paramagnetic

Shift Reagents.

Z. W. Wolkowski, Tetrahedron Lett. 821 (1971).
Tris(dipivalomethanato)Ytterbium-Induced Shifts in PMR
Ketones and Aldehydes.

Z. W. Wolkowski, Tetrahedron Lett. 825 (1971). Tris
(dipivalomethanato)Europium-Induced Shifts in PMR
Oximes.

Z. W. Wolkowski, J. Cassan, L. Elegant and M. Azzaro,
C. R. Acad. Sci., Paris, Ser. C, <u>272</u>, 1244 (1971).
Étude par Résonance· Magnetique du Proton de Quelques
Oximes Terpéniques.

Z. W. Wolkowski, C. Beaute and R. Jantzen, Chem.
Commun. 619 (1972). Shifts Induced by Tris(dipivalomethanato)lanthanides in ^{19}F NMR Spectroscopy.

H. Yanagawa, T. Kato and Y. Kitahara, Tetrahedron
Lett. 1073 (1973). Asparagusic Acid-S-Oxides, New
Plant Growth Regulators in Etiolated Young Asparagus
Shoots.

K. C. Yee and W. G. Bentrude, Tetrahedron Lett. 2775
(1971). Six-Membered Ring Phosphorus Heterocycles.
Use of Europium Shift Reagent in the Analysis of the
PMR. Spectrum of Trans -2-Methyl-5-t-Butyl-2-Oxo-1,
3, 2-Dioxaphosphorinane.

M. Yoshimoto, T. Hiraoka, H. Kuwano and Y. Kishida,
Chem. Pharm. Bull. (Jap.), <u>19</u>, 849 (1971). Use of a
Shift Reagent in First-Order Analysis of Cyclopropane
Derivatives in NMR Spectroscopy.

E. E. Zaev V. K. Voronov, M. S. Shvartsberg, S. F.
Vasilevsky, Yu. N. Molin and I. L. Kotliarevsky. Tetra-
hedron Lett. 617 (1968). Application of Paramagnetic
Additions to the Structure Determination of Some

Pyrazoles by NMR.

APPLICATIONS OF SHIFT REAGENTS TO STUDIES OF
BIOLOGICALLY SIGNIFICANT MOLECULES

N.S. Angerman and S.S. Danyluk, Abstracts 13th Experimental NMR Conference, Pacific Grove, Calif., May 1972, p. 51. The Conformation of Chloroquine, an Antimalarial, in Solution Using a Shift Reagent.

N.S. Angerman, S.S. Danyluk and T.A. Victor, J. Amer. Chem. Soc., 94, 7137 (1972). A Direct Determination of the Spatial Geometry of Molecules in Solution. I. Conformation of Chloroquine, an Antimalarial.

C.D. Barry, J.A. Glasel, A.C.T. North, R.J.P. Williams and A.V. Xavier, "Proceedings of the 9th Rare Earth Research Conference", Ed. by P.E. Field, Blacksburg, Va., Oct 10 - 14, 1971, p. 263. The Lanthanide Cations as NMR Conformational Probes.

C.D. Barry, J.A. Glasel. A.C.T. North. R.J.P. Williams and A.V. Xavier, Biochim. Biophys. Acta, 262, 101 (1972). The Quantitative Conformations of Some Dinucleoside Phosphates in Solution.

C.D. Barry, J.A. Glasel, A.C.T. North, R.J.P. Williams and A.V. Xavier, Biochem. Biphys. Res. Commun., 47, 166 (1972). The Conformations of Adenosine Mononucleotide in Water and Dimethylsulfoxide.

C.D. Barry, A.C.T. North, J.A. Galsel, R.J.P. Williams and A.V. Xavier, Nature, 232, 236 (1971). Quantitative Determination of Mononucleotide Conformations in Solution Using Lanthanide Ion Shift and Broadening NMR Probes.

E. Bayer and K. Beyer, Tetrahedron Lett. 1209 (1973). Signalzuordnung in ^{13}C-NMR-Spektren von Oligopeptiden

389

und Sequenzbestimmung mit Hilfe von $Pr(ClO_4)_3$ als Verschiebungsreagens.

E. R. Birnbaum, C. Yoshida, J. E. Gomez and D. W. Darnall, "Proceedings of the 9th Rare Earth Research Conference", Ed. by P. E. Field, Blacksburg, Va., Oct 10-14, 1971, p. 264. Investigations of Amino Acid Complexes of Nd(III).

D. W. Darnall, E. R. Birnbaum, J. E. Gomez and G. E. Smolka, Proceedings of the 9th Rare Earth Research Conference, Ed. by P. E. Field, Blacksburg, Va., Oct. 10-14, 1971, p. 278. Rare Earth Metal Ions as Probes of Calcium Ion Binding Sites in Proteins.

E. W. Darnal and E. R. Birnbaum, J. Biol. Chem., 245, 6484 (1970). Rare Earth Metal Ions as Probes of Calcium Ion Binding Sites in Proteins.

R. A. Dwek, K. G. Moralee, R. E. Richards, R. J. P. Williams and A. V. Xavier, "Proceedings of the 9th Rare Earth Research Conference", Ed. by P. E. Field, Blacksburg, Va., Oct 10-14, 1971, p. 518. Molecular Conformation Determinations of Inhibitor/Enzyme Complexes with Respect to the Gd(III) Reporter Site.

R. A. Dwek, R. E. Richards, K. G. Moralee, E. Nieboer, R. J. P. Williams and A. V. Xavier, European J. Biochem., 21, 204 (1971). The Lanthanide Cations as Probes in Biological Systems. Proton Relaxation Enhancement Studies for Model Systems and Lysozyme.

P. Girard, H. Kagan and S. David, Tetrahedron, 27, 5911 (1971). Spectres de RMN de Quelques Derives de Sucres en Presence de Chelates de Terres Rares.

J. A. Glasel, R. J. P. Williams and A. Xavier, Abstracts, 13th Experimental NMR Conference, Pacific Grove, Calif., May 1972, p. 7. Lanthanide ions as Probes of

Structures of Biological Importance in Polar Solvents.

R. J. Kostelnik and S. M. Castellano, Abstracts, 13th
Experimental NMR Conference, Pacific Grove, Calif.,
May 1972, 7. 250 MHz PMR Spectrum of a Sonicated
Lecithin Dispersion in Water. The Effect of Ferri-
cyanide, Maganese(II), Europium(III), and Gadolinium
(III) Ions on the Choline Methyl Resonance.

G. Lukacs, Bull. Soc. Chim. France, 350 (1972).
Utilisation de Precurseurs Enrichis en ^{13}C dans l'
Etudes par la RMN de Problemes Biosynthetiques.

K. G. Morallee, E. Nieboer, F. J. C. Rossotti, R. J. P.
Williams, A. V. Xavier and R. A. Dwek, Chem. Commun.
1132 (1970). The Lanthanide Cations as NMR Probes
of Biological Systems.

E. Nieboer, "Proceedings of the 9th Rare Earth Research Conference, Ed. by P. E. Field, Blacksburg, Va.,
Oct 10-14, 1971, p. 262. The Lanthanide Cations as
NMR Probes of Biological Systems: Studies of Lysozyme
and Staphylococcal Nuclease.

J. Reuben, "Proceedings of the 9th Rare Earth Research
Conference", Ed. by P. E. Field, Blacksburg, Va., Oct.
10-14, 1971, p. 514. Gadolinium(III) as a Paramagnetic
Probe for Magnetic Resonance Studies of Biological
Macromolecules.

A. D. Sherry, C. Yoshida, E. R. Birnbaum and D. W.
Darnall, J. Amer. Chem. Soc., 95, 3011 (1973). NMR
Study of the Interaction of Nd(III) with Amino Acids and
Carboxylic Acids. An Aqueous Shift Reagent.

H. H. Strain, W. A. Svec, K. Aitzetmuller, M. C.
Grandolfo, J. J. Katz, H. Kjøsen, S. Norgard, S.
Liaaen-Jensen, F. T. Haxo, P. Wegfahrt and H.
Rapoport, J. Amer. Chem. Soc., 93, 1823 (1971). The

Structure of Peridinin, the Characteristic Dinoflagellate Carotenoid.

W. Voelter, C. Bürvenich and E. Breitmaier, Angew. Chem. (Intern. Edit.), 11, 539 (1972). Complex Shift Reagent for ^{13}C NMR Studies on Carbohydrates.

R.J.P. Williams, Quart. Rev. (London), 24, 331 (1970). The Biochemistry of Sodium, Potassium, Magnesium, and Calcium.

DETERMINATION OF ENANTIOMERIC AND DIASTEREOMERIC COMPOSITIONS AND RELATED PHENOMENA

J.T. Blackwell and F.I. Carroll, 24th Southeastern Regional ACS Meeting, Birmingham, Alabama, Nov. 2, 1972. The Determination of the Optical Purity of Amino Alcohols.

A.K. Bose, B. Dayal, H.P.S. Chawla and M.S. Manhas, Abstract, PHYS 73, 164th ACS National Meeting, New York, N.Y., Aug. 27-Sept. 1, 1972. Effect of Shift Reagents on Diastereotopic Protons in Some β-Lactams.

A.K. Bose, B. Dayal, H.P.S. Chawla and M.S. Manhas, Tetrahedron Lett. 3599 (1972). Unexpected Effect of Shift Reagents on Diastereotopic Protons in Some β-Lactams.

E.B. Dongala, A. Solladie-Cavallo and G. Solladie, Tetrahedron Lett., 41, 4233 (1972). Determination de la Purete Enantiomerique de β-Hydroxyesters Partiel-lement Actifs Par RMN. Utilisation Du Tris(Trifluoro-methylhydroxymethylene)-3 camphorato-d-europium(III).

J.N. Eikenberry, Ph.D. Thesis (Part I), "Fluorinated Acyl Derivatives of d-Camphor Chelated to Lanthanides: Chiral Shift Reagents," University of Wisconsin,

Madison, Wisconsin, <u>1972.</u>

R. R. Fraser, M. A. Petit and M. Miskow, J. Amer. Chem. Soc., <u>94</u>, 3253 (1972). Separation of NMR Signals of Internally Enantiotropic Protons Using a Chiral Shift Reagent. The Deuterium Isotope Effect on Geminal Proton -Proton Coupling Constants.

R. R. Fraser, M. A. Petit and J. K. Saunders, Chem. Commun. 1450 (1971). Determination of Enantiomeric Purity by an Optically Active NMR Shift Reagent of Wide Applicability.

H. Gerlach and B. Zagalak, Chem. Commun. 274 (1973). Determination of the Enantiomeric Purity and Absolute Configuration of α-Deuteriated Primary Alcohols.

H. L. Goering, J. N. Eikenberry and G. S. Koermer, J. Amer. Chem. Soc., <u>93</u>, 5913 (1971). Tris(3-(trifluoro-methylhydroxymethylene)-<u>d</u>-camphorato)europium(III). A Chiral Shift Reagent for Direct Determination of Enantiomeric Compositions.

J. L. Green, Jr., and P. B. Shelvin, Chem. Commun. 1092 (1971). <u>Meso</u> - and (<u>+</u>)-Bis(phenylsulphinyl)methane. Characterization Using NMR Chemical Shift Reagents.

P. Joseph-Nathan, J. E. Herz and V. M. Rodriguez, Can. J. Chem., <u>50</u>, 2788 (1972). Doubling Proton Magnetic Resonance Signals by Use of Tris(dipivalomethanato)-europium(III).

M. Kainosho, K. Ajisaka, W. Pirkle and S. D. Beare, J. Amer. Chem. Soc., <u>94</u>, 5924 (1972). The Use of Chiral Solvents or Lanthanide Shift Reagents to Distinguish Meso from <u>d</u> or <u>l</u> Diastereomers.

K. Nakanishi and J. Dillion, J. Amer. Chem. Soc., <u>93</u>, 4058 (1971). A Simple Method for Determining the

Chirality of Cyclic α -Glycols with $Pr(dpm)_3$ and $Eu(dpm)_3$.

K. Nakanishi, D. A. Schooley, M. Koreeda and J. Dillon, Chem. Commun. 1235 (1971). Absolute Configuration of of the C_{18} Juvenile Hormone: Application of a New Circular Dichroism Method Using Tris(dipivaloylmethanato)-praseodymium.

H. Nozaki, K. Yoshino, K. Oshima and Y. Yamamoto, Bull. Chem. Soc. Jap., 45, 3495 (1972). The Determination of the Enantiomeric Purity of Methyl p-Tolyl Sulfoxide by Means of an NMR Shift Reagent.

W. H. Pirkle, T. G. Burlingame and S. D. Beare, Tetrahedron Lett. 5849 (1968). Optically Active NMR Solvents VI. The Determination of Optical Purity and Absolute Configuration of Amines.

A. Rahm and M. Pereyre. Tetrahedron Lett. 1333 (1973). Détermination de la Pureté Enantiomérique d'un Carbure Organostannique au Moyen de la RMN: Utilisation des Complexes d'europium de Composés Organostanniques Functionnels.

G. P. Schiemenz, Tetrahedron Lett. 4267 (1972). Zur Wirkung von Verschiebungsreagentien auf Diastereotope Protonen in β -Lactamen.

G. P. Schiemenz and H. Rast, Tetrahedron Lett. 4685 (1971). Lanthanide Verschiebungsreagentien als Hilfsmittel bei Diastereatropie Problemen.

G. P. Schiemenz and H. Rast, Tetrahedron Lett. 1697 (1972). Isochronie Heterotoper Kerne.

V. Schurig, Inorg. Chem., 11, 736 (1972). Chiral d^8 Metal Ion Coordination Compounds. The Preparation of d-3-Trifluoroacetylcamphorato Complexes of Rhodium,

Palladium, and Nickel.

V. Shurig, Tetrahedron Lett. 3297 (1972). Chiral Shift
Reagents for NMR Spectroscopy. A Simple and Improved
Access to Lanthanide-Tris-Chelates of d-3-TFA-
Camphor.

G. M. Whitesides and D. W. Lewis, J. Amer. Chem.
Soc., 92, 6979 (1970). Tris (3-tert-butylhydroxy-
methylene)-d-camphorato)europium(III). A Reagent
for Determing Enantiomeric Purity.

G. M. Whitesides and D. W. Lewis, J. Amer. Chem.
Soc., 93, 5914 (1971). The Determination of Enantiomeric Purity Using Chiral Lanthanide Shift Reagents.

G. E. Wright, Tetrahedron Lett. 1097 (1973). NMR
Assignment of Diastereotropic Protons in Amphetamine.

KINETIC STUDIES INVOLVING
SHIFT REAGENTS

K. D. Berlin and S. Rengaraju, Abstracts, 163rd Na-
tional ACS Meeting Boston, April 1972. A Case of Slow
Nitrogen Inversion Due to Intramolecular Hydrogen
Bonding. Study of Slow Nitrogen Inversion in Diethyl 2-
Aziridinylphosphonate from the Paramagnetic-Induced
Shifts in the PMR Spectra Using Tris(dipivalomethanato)-
europium(III).

H. N. Cheng and H. S. Gutowsky, J. Amer. Chem. Soc.,
94, 5505 (1972). The Use of Shift Reagents in NMR
Studies of Chemical Exchange.

D. F. Evans and M. Wyatt, Chem. Commun. 312 (1972).
Direct Observation of Free and Complexed Substrate in
a Lanthanide Shift Reagent System.

F. A. Hart, J. E. Newbery and D. Shaw, J. Inorg. Nucl.

Chem., 32, 3585 (1970). Lanthanide Complexes-X. PMR Studies of Alkyl-Substituted Bipyridine Complexes of Lanthanides: Paramagnetic Shifts and Reaction Kinetics.

Y. Takagi, S. Teratani and J. Uzawa, Chem. Commun. 280 (1972). Application of NMR Shift Reagents to Kinetic Studies on Catalytic Deuteration of 4-t-Butylcyclohexanone.

ISOTOPE EFFECTS IN SHIFT REAGENT CHEMISTRY

J.J. Brooks and R.E. Sievers, J. Chromatographic Sci., 11, in press (1973). Gas Chromatographic Studies of the Interactions Between Selected Organic Nucleophiles and the NMR Shift Reagent, Tris(1, 1, 1, 2, 2, 3, 3-heptafluoro-7, 7-dimethyl-4, 6-octanedionato)europium (III).

R.R. Fraser, M.A. Petit and M. Miskow, J. Amer. Chem. Soc., 94, 3253 (1972). Separation of NMR Signals of Internally Enantiotropic Protons Using a Chiral Shift Reagent. The Deuterium Isotope Effect on Geminal Proton-Proton Coupling Constants.

A.M. Grotens, C.W. Hilbers and E. de Boer, Tetrahedron Lett. 2067 (1972). Isotope Effects in NMR Spectra of Polyglycoldimethylethers Complexed with Rare-Earth Shift Reagents.

C.C. Hinckley, W.A. Boyd and G.V. Smith, Tetrahedron Lett. 879 (1972). Deuterium Isotope Effects Observed in NMR Shifts Induced by Lanthanide Complexes.

J.J.M. Sanders and D.H. Williams, Chem. Commun. 436 (1972). Secondary Deuterium Isotope Effect on Lewis Basicity: Tris (dipivaloylmethanato)europium as a Simple and Effective Probe.

G. V. Smith, W. A. Boyd and C. C. Hinckley, J. Amer. Chem. Soc., 93, 6319 (1971). Isotope Effects in NMR Spectra Modified by Rare-Earth Shift Reagents.

APPLICATIONS OF SHIFT REAGENTS TO THE STUDY OF STEREOCHEMISTRY

L. J. Altman, R. C. Kowerski and H. C. Rilling, J. Amer. Chem. Soc., 93, 1782 (1971). Synthesis and Conversion of Presqualene Alcohol to Squalene.

R. von Ammon, R. D. Fischer and B. Kanellakopulos, Chem. Ber., 104, 1072 (1971). ¹H-NMR-Utersuchungen der Konformativen Beweglichkeit von Monosubstituiertem Cyclohexan mit Hilfe eines Paramagnetischer Seitenerd-Ions im Substitutenten: Triscyclopentadienyl(cyclohexylisonitril)-praseodym(III).

N. H. Andersen, H. Uh, S. E. Smith and P. G. M. Wuts, Chem. Commun. 956 (1972). Stereochemical Course of the Cyclization of Olefinic Aldehydes.

J. W. ApSimon and J. D. Cooney, Can. J. Chem., 49, 2377 (1971). Structure and NMR Spectrum of N-Nitrosocamphidine.

I. M. Armitage, L. D. Hall, A. G. Marshall and L. G. Werbelow, Abstract ORGN 146, 164th ACS National Meeting, New York, N. Y., Aug 27-Sept. 1, 1972. Use of Lanthanide NMR Shift Reagents in Determination of Molecular Configuration.

I. M. Armitage, L. D. Hall, A. G. Marshall and L. G. Werbelow, J. Amer. Chem. Soc., 95, 1437 (1973). Use of Lanthanide NMR Shift Reagents in Determination of Molecular Configuration.

J. Baciszewski, A. J. Rafalski and M. Wiewiorowski, Bull. Acad. Pol. Sci., Ser. Sci. Chim., 19, 545 (1971).

397

Europium Shifts in the NMR Spectra of Lactams. I.
Stereochemical Assignment of Lupanine.

R. Bauman, Tetrahedron Lett. 419 (1971). Differentiation of cis and trans Isomers of Thionocarbamate
Esters by use of Tris(Dipivalomethanato)europium(III).

C. Beaute, Z.W. Wolkowski, J.P. Merda and D.
Lelandais, Tetrahedron Lett. 2473 (1971). Structural
Assignment of Chloro-and Bromo-Vinyl Aldehydes by
Yb(dpm)$_3$ -Assisted Proton NMR .

P. Bélanger, C. Freppel, D. Tizané and J.C. Richer,
Can. J. Chem., 49, 1985 (1971). Deplacement Induit par
un Lanthanide en RMN. Une Application a un Probleme
Stereochimique.

W.G. Bentrude, H. -W. Tan and K.C. Yee, J. Amer.
Chem. Soc., 94, 3264 (1972). Conformations of Saturated Phosphorus Heterocycles. Effect of Europium
Dipivaloylmethane and Europium Heptafluorodimethyl-
octanedione on Conformational Equilibria of 2-Substituted
5-tert -Butyl-2-oxo-1, 3, 2-dioxaphosphorinanes.

K.D. Berlin and S. Rengaraju, J. Org. Chem. 36, 2912
(1971). A Study of Syn/Anti Oxime Ratios from the Paramagnetic-Induced Shifts in the PMR Spectra Using Tris-
(dipivalomethanato)europium(III).

J.A. Berson, R.T. Luibrand, N. G Kundu and D.G.
Morris, J. Amer. Chem. Soc., 93, 3075 (1971).
Carbonium Ion Rearrangements of Bicyclo (2.2.2)oct-
2-ylcarbinyl Derivatives.

G. Borgen, Acta Chem. Scand., 26, 1740 (1972). The
Conformation of 4, 4, 7, 7-Tetramethylcyclononanone;
Low-Temperature NMR ·Spectroscopy in Conjunction
with the Shift Reagent Eu(dpm)$_3$.

H. van Brederode and W. G. B. Huysmans, Tetrahedron Lett. 1695 (1971). Analysis of Diamine Stereoisomers from Proton Paramagnetic Shifts.

E. Brown, R. Dhial and P. F. Casals, Tetrahedron, 28, 5607 (1972). Stereochimie des Dialcoyl-2, 6 Piperidinols-3.

J. F. Caputo and A. R. Martin, Abstracts, 163rd National ACS Meeting, Boston, April 1972. 1, 4-Benzodioxans and 1, 4-Benzorathians. 2. Use of Lanthanide NMR Shift Reagents for Stereochemical Structure Elucidation.

J. F. Caputo and A. R. Martin, Tetrahedron Lett. 4547 (1971). Stereochemical Elucidation of Isomeric Tricyclic 1, 4-Benzodioxans (II) by the Use of the NMR Shift Reagent, Eu(fod)$_3$.

F. A. Carey, J. Org. Chem., 36, 2199 (1971). Application of Europium(III) Chelate-Induced Chemical Shifts to Stereochemical Assignments of Isomeric Perhydrophenalenols.

F. I. Carroll and J. T. Blackwell, Tetrahedron Lett. 4173 (1970). Structure and Conformation of cis and trans-3, 5-Dimethylvalerolactones.

C. P. Casey and R. A. Boggs, Tetrahedron Lett. 2455 (1971). Stereospecific Addition of Lithium Dipropenylcuprate to 2-Cyclohexenone.

J. R. Corfield and S. Trippett, Chem. Commun. 721 (1971). Assignment of Configuration to 2, 2, 3, 4, 4-Pentamethylphosphetan Oxides Using Tris(dipivalomethanato)europium(III).

L. Crombie, D.A.R. Findley and D.A. Whiting, Tetra-
hedron Lett. 4027 (1972). Lanthanide-Induced Chemical
Shifts in Natural Cyclopropanes: Stereochemistry of
Chrysanthemy and Presqualene Alcohols and Esters.

L. Ernst, Chem. -Ztg. Chem. App., 95, 325 (1971).
Zur Kernresonanz-Spektroskopie mit Lanthaniden-
Komplexen: Konfigurationsbestimmung der Isomeren
1, 3-Dimethylcyclohexylamine-(2) mit Hilf von Tris-
(dipivaloylmethanato)europium(III).

R.R. Fraser and Y.Y. Wigfield, Tetrahedron Lett. 2515
(1971). Effects of Stereochemistry on the Stability of
Alpha-Nitrosamino Carbanions.

P. Granger, M.M. Claudon and J.F. Guinet, Tetrahedron Lett. 4167 (1971). Utilisation du Tris(dipivalomethanato)europium(III) en Etudes Confromationnelles
cas des α-Monobenzyl-cyclohexanols cis et trans.

J.W. de Haan and L.J.M. van de Ven, Tetrahedron Lett.
2703 (1971). Z-E Conformational Isomerism of Nerol,
Geraniol and their Acetates.

H. Hogeveen, C.F. Roobeek and H.C. Volger, Tetrahedron Lett. 221 (1972). Inversion of Configuration in a
Fused Cyclopropane Ring Opening by Hydrochloric Acid.

D. Horton and J.K. Thomson, Chem. Commun. 1389
(1971). Application of a Lanthanide Shift Reagent for
Conformational and Configurational Assignment in the
Carbohydrate Field.

R.O. Hutchins and B.E. Maryanoff, Abstract ORGN 58,
164th ACS National Meeting, New York, N Y., Aug. 27-
Sept. 1, 1972. The Use of Eu(fod)$_3$ Shift Reagent as a
Configurational and Conformational Probe in 2-R-2-Oxo-
2-phospha-1, 3-dithiocyclohexanes.

S. Itô and I. Itoh, Tetrahedron Lett. 2969 (1971). Cyclo-addition Reaction of Tropylium Ion and Cyclopentadiene.

L. F. Johnson, J. Chakravarty, R. Dasgupta and U. R. Ghatak, Tetrahedron Lett. 1703 (1971). Application in the Use of a Shift Reagent in PMR Spectroscopy for the Elucidation of Structures and Stereochemistry of Epimeric Methyl 4, 9-Dimethyl-7, 8-Benzobicyclo (3.3.1) Non-7-ene-4-Carboxylates.

A. Kato and M. Numata, Tetrahedron Lett. 203 (1972). Brugierol and Isobrugierol trans-and cis-1, 2-dithiolane-1-oxide, from Brugiera Conjugata.

H. Kessler and D. Rosenthal, Tetrahedron Lett. 393 (1973). Z, E-isomerie Aromatischer Diazoketone.

M. Kishi, K. Tori, T. Komeno and T. Shingu, Tetrahedron Lett. 3525 (1971). Application of Paramagnetic Shift Induced by Tris(dipivalomethanato)europium(III) to Configurational Assignment of Sulfinyl Oxygen in 5α-Cholestan-2α, 5-Episulfoxides. Examples of Upfield Shifts Due to Eu(dpm)$_3$.

R. W. Kreilick, J. Amer. Chem. Soc., 90, 5991 (972). NMR Studies of Phenoxy Radicals. Spin Delocalization in Cyclic Aliphatic Substituents.

L. Lacombe, F. Khuong-Huu, A. Pancrazi, Q. Khuong-Huu and G. Lukacs, C. R. Acad. Sci., Paris, Ser. C, 272, 668 (1971). Steroid Alkaloids. CXIX. Study of NMR Spectra of Epimeric Aminosteroids in the Presence of Eu(dpm)$_3$.

S. G. Levine and R. E. Hicks, Tetrahedron Lett. 311 (1971) The Conformation of Griseofulvin. Application of an NMR Shift Reagent.

G. J. Martin, N. Naulet, F. Lefevre and M. L. Martin, Org. Mag. Res., 4, 121 (1972). Determination Rapide de la Configuration cis - trans des Alcenes CH-CH= CH Utilisation des Complexes de Terres Rares.

J. D. McKinney, L. H. Keith, A. Alford and C. E. Fletcher, Can. J. Chem., 49, 1993 (1971). The PMR Spectra of some Chlorinated Polycyclodiene Pesticide Metabolites. Rapid Assessment of Stereochemistry.

G. Montuado and P. Finnocchiaro, J. Org. Chem., 37, 3434 (1972). Determination of the Molecular Geometry of Eu(fod)$_3$ Complexes with Amides and Diamides and its Conformational Significance.

G. F. Morris, H. R. Zandstra, 24th Southeastern Regional ACS Meeting, Birmingham, Alabama, Nov. 2, 1972. Configurational Proof of 7-Methylbicyclo (4.2.1) Nonan-9-ones: Use of NMR Shift Reagent, Eu(fod)$_3$.

G. A. Neville, Org. Mag. Res., 4, 633 (1972). Characterization of Di- and Trimethoxybenzaldehydes by Analysis of Deshielding Gradients Obtained with Eu(fod)$_3$ Shift Reagent.

M. Ochiai, E. Mizuta, O. Aki, A. Morimoto and T. Okada, Tetrahedron Lett. 3245 (1972). The Determination of Stereochemistry of 3-Methylenecepham Derivatives by Means of Lanthanide-Induced Shifts.

M. Ohashi, I. Morishima and T. Yonezawa, Bull. Chem. Soc. Jap., 44, 576 (1971). Application of Proton NMR Shift Reagents to the Stereochemical Analysis of Nicotine.

T. Okutani, A. Morimoto, T. Kaneko and D. Masuda, Tetrahedron Lett. 1115 (1971). Studies on Azetidine Derivatives III. Configurational Assignment with a Shift Reagent.

L. A. Paquette, R. S. Beckley, D. Truesdell and J. Clardy, Tetrahedron Lett. 4913 (1972). Substitutent Control of Stereochemistry in the Rearrangements of 1, 8-Bishomocubanes Catalyzed by Silver (I) Ion.

L. A. Paquette, S. A. Lang. Jr., S. K. Porter and J. Clardy, Tetrahedron Lett. 3137 (1972). Hydration of Hexamethyl(dewarbenzene)oxide. Stereochemical Aspects öf the Resulting Rearrangement.

J. A. Peters, J. D. Remijnse, A. van der Wiele and H. van Bekkum, Tetrahedron Lett. 3065 (1971). Synthesis and (Non-Chair) Conformation of some 3α, 7α-Disubstituted Bicyclo (3.3.1.) Nonanes.

D. C. Remy and W. A. Van Saun, Jr., Tetrahedron Lett. 2463 (1971). Determination of the Stereochemical Configuration of 3-Substituted N, N-Dimethyl-5H-Dibenzo-(a, d) Cycloheptene -$\Delta^{5, \gamma}$ Propylamine Derivatives Using Tris(dipivalomethanato)-Europium as a Shift Reagent.

A. J. M. Reuvers, A. Sinnema and H. van Bekkum, Tetrahedron, 28, 4353 (1972). Configurational Analysis of 4,7,7-trimethyltricyclo (2.2.1. O$^{4, 6}$)heptan-3-ols Using Eu(dpm)$_3$.

H. G. Richey, Jr., and F. W. Von Reing, Tetrahedron Lett. 3781 (1971). Determination of the Configurations (cis or trans) of Alkenols by the Effects of Paramagnetic Shift Reagents on their NMR Spectra.

A. A. M. Roof, A. van Wageningen, C. Kruk and H. Cerfontain, Tetrahedron Lett. 367 (1972). Photochemistry of Non-conjugated Dienones II: Electrocyclic Reactions of Some 1, 2-Dimethylenecyclohexanes.

J. R. Salaün and J. M. Conia, Tetrahedron Lett. 2849 (1972). The Stereospecific Thermal Ring Enlargement of 1-Vinylcyclopropanels into Cyclobutanone Derivatives.

H. -D. Scharf and M.-H. Feilen, Tetrahedron Lett. 2745 (1971). Strukturaufklarung Eines Polyfunktionellen Isomerenpaares mit Hilfe des Tris(dipivaloylmethanato)-Europium Komplexes.

E. C. Sen and R. A. Jones, Tetrahedron, 28, 2871 (1972). The Chemistry of Terpenes - IV. The Configuration and Conformation of the Isomeric 3-Aminopinanes.

T. Shingu, T. Hayashi and H. Inouye, Tetrahedron Lett. 3619 (1971). Zur Stereochemie des Cataponols, Ein Beispiel der Anwendung des Verschiebungs-Reagenzes zur Konfigurationsaufklarung.

K. -E. Stensio and U. Åhlin, Tetrahedron Lett. 4729 (1971). Synthesis and Europium-Shifted NMR Spectra of trans- and cis-4 (β (1-Naphthyl)Vinyl) Pyridine.

A. Tangerman and B. Zwaneburg, Tetrahedron Lett. 79 (1972). Assignment of Configuration to Sulfines by Means of Shift Reagents and Asis.

J. M. Tronchet, F. Barbalat-Rey and N. Le-Hong, Helv. Chim. Acta, 54, 2615 (1971). Equilibres Conformationnels de Glucides au Niveau de Liaisons sp^2 - sp^3 C-C. III. Derives d'Oximes d'Aldehydo-sucres Utilisation de Tris-dipivaloylmethanato Europium.

M. R. Vegar and R. J. Wells, Tetrahedron Lett. 2847 (1971). The Conformational Analysis of Some 3-Substituted Bicyclo (3.3.1) -Nonanes by Means of $Eu(dpm)_3$-Induced NMR Shifts.

W. Walter, R. F. Becker and J. Thiem, Tetrahedron Lett. 1971 (1971). Configurational Assignment of Thioamides Using Pseudocontact Shifts Induced by $Eu(dpm)_3$

R. Wasylishen and T. Schaefer, Can. J. Chem., 50, 274 (1972). PMR Spectra Conformational Preferences, and Approximate Molecular Orbital Calculations for the syn and anti 2- Furanaldoximes.

G. M. Whitesides and J. San Filippo, Jr., J. Amer. Chem. Soc., 92, 6611 (1970). The Mechanism of Reduction of Alkylmercuric Halides by Metal Hydrides.

M. R. Willcott, J. F. M. Oth, J. Thio, G. Plinke and G. Schroder, Tetrahedron Lett. 1579 (1971). Rapid Solution of Stereochemical Problems with Europium-Tris(Tetramethylheptanedione), $Eu(thd)_3$.

J. deWitt and H. Wynberg, Spectroscopy Lett., 5, 119 (1972). Configuration of a Naphtho (2, 3-B)Thiophene-Diels-Alder Adduct from NMR Shift Reagent Data.

G. Wood, G. W. Buchanan and M. H. Miskow, Can. J. Chem., 50, 521 (1972). Conformational Studies of Substituted Trimethylene Sulfites by Proton Magnetic Resonance.

GLOSSARY

A: lanthanide shift reagent

acac: 2,4-pentanedionate

acam: 3-acetyl-d-camphorate

B: Lewis base

BPO: dibenzoyl peroxide

BPPO: benzoyl propionyl peroxide

C_+: chiral shift reagent (where + denotes one optical isomer of the chelate)

CBPPO: m-chlorobenzoyl propionyl peroxide

CIDNP: chemically-induced dynamic nuclear polarization

CP: cyclopentanone

D: donor, Lewis base

dbm: 1,3-diphenyl-1,3-propanedionate

dfhd: 1,1,1,5,5,6,6,7,7,7-decafluoro-2,4-heptanedionate (referred to in previous literature as 1,1,1,2,2,3,3,-7,7,7-decafluoro-4,6-heptanedionate; the present numbering system is that used in Chemical Abstracts)

dibm: 2,6-dimethyl-3,5-heptanedionate

DMF: N,N-dimethylformamide

dmp: 5,6-dimethyl-1,10-phenanthroline

DMSO: dimethylsulfoxide

DNMR: dynamic nuclear magnetic resonance spectroscopy

dpm, DPM: 2,2,6,6-tetramethyl-3,5-heptanedionate (synonymous with thd and tmhd)

dpmH: 2,2,6,6-tetramethyl-3,5-heptanedione or dipivaloyl-methane

E: lanthanide shift reagent

E.P.: enantiomeric purity

facam: 3-trifluoroacetyl-d-camphorate (synonymous with 3-trifluoromethylhydroxymethylene-d-camphorate)

fhd: 1,1,1,2,2,6,6,7,7,7-decafluoro-3,5-heptanedionate

fod, FOD: 6,6,7,7,8,8,8-heptafluoro-2,2-dimethyl-3,5-octanedionate (referred to in previous literature as 1,1,1,2,2,3,3-heptafluoro-7,7-dimethyl-4,6-octanedionate; the present numbering system is that used in Chemical Abstracts)

FOZ: first-order Zeeman

H(fod): 6,6,7,7,8,8,8-heptafluoro-2,2-dimethyl-3,5-octanedione.

I: impurity

L: lanthanide element

LAN: lanthanide element

LIS: lanthanide-induced shift

Ln: lanthanide element

Ln(fod)$_3$: 6, 6, 7, 7, 8, 8, 8-heptafluoro-2, 2-dimethyl-3, 5-octanedionate of a lanthanide

Ln(thd)$_3$: 2, 2, 6, 6-tetramethyl-3, 5-heptanedionate of a lanthanide (synonymous with Ln(DPM)$_3$)

LSR: lanthanide shift reagent

MO: molecular orbital

n_R: mole fraction of R enantiomer

n_S: mole fraction of S enantiomer

phen: 1, 10-phenanthroline

pta: 1, 1, 1-trifluoro-5, 5-dimethyl-2, 4-hexanedionate

Py: pyridine

R: (Chapter 4) R enantiomer
(Chapter 7) agreement factor
(Chapters 9 and 15) distance
(Chapter 16) shift reagent

S: substrate (Lewis base)
(Chapter 4 only) S enantiomer

SOZ: second-order Zeeman

THF: tetrahydrofuran

thd: 2, 2, 6, 6-tetramethyl-3, 5-heptanedionate

TMC: trimethylcarbamate

tmhd: 2, 2, 6, 6-tetramethyl-3, 5-heptanedionate

TMS: tetramethylsilane

X: donor atom of a substrate

NOTICE

The authors and publisher do not make any warranty, express or implied, or assume any legal liability or responsibility for the accuracy, completeness, or usefulness of any information, apparatus, product or process disclosed, or represent that its use would not infringe privately owned rights. Shift reagent compositions and their uses in NMR spectroscopy are the subjects of U.S. Patent 3,700,410 and pending foreign and U.S. applications. Eastman Kodak, Rochester, New York and Willow Brook Laboratories, Waukesha, Wisconsin have been licensed under these patent rights and the compounds are commercially available from them.